Hands-on Organic Chemistry

Hands-on Organic Chemistry

A Visual Laboratory Guide for Students and Professionals

Thomas Norberg

WILEY VCH

Author

Prof. Dr. Thomas Norberg
Department of Chemistry,
Uppsala University,
Uppsala, Sweden

Cover Design: Wiley
Cover Image: Thomas Norberg

Typesetting: Lumina Datamatics, Madurai, India

Library of Congress Card No.: applied for

British Library Cataloguing-in-Publication Data A catalogue record for this book is available from the British Library.

Bibliographic information published by the Deutsche Nationalbibliothek The Deutsche Nationalbibliothek lists this publication in the Deutsche Nationalbibliografie; detailed bibliographic data are available on the Internet at <http://dnb.d-nb.de>.

Print ISBN: 9783527356683
ePDF ISBN: 9783527857180
ePub ISBN: 9783527857173

Druck und Bindung: CPI Group (UK) Ltd, Croydon, CR0 4YY

C9783527356683_270726

Contents

Preface

Many good books about practical organic chemistry have been published. I regard "Vogel's Textbook of Practical Organic Chemistry," first published in 1948, as the golden standard. The latest version (5th edition, from 1989) has 1500+ pages and is still unsurpassed in scope and size. Some later books about practical organic chemistry are referenced in the first three chapters. With a few exceptions, Leonard et al.'s "Advanced Practical Organic Chemistry" being the most obvious, these books are organized in the following way: basic organic chemistry laboratory techniques are briefly described, followed by a large section containing preparations that illustrate different reaction classes and lab techniques. The primary aim of these books is to help *teachers* of organic chemistry to organize a meaningful laboratory course. The primary aim of this book, on the other hand, is to help *students* at all levels, from undergraduate to PhD students to postdocs, to understand and master the preparative aspects of organic chemistry. Basic organic chemistry lab techniques (milligram-to-gram scale) are described in detail using a picture-driven approach. The focus is on *simple* equipment, step-by-step descriptions, and useful tips.

Much practical lab information can nowadays be gathered from videos, pictures, and texts present on the internet, but it is my impression that this information is very scattered and that the quality varies. As regards the videos, they are very efficient in presenting equipment and procedures, but less efficient when it comes to explaining the background and theory of lab procedures. Videos can also suffer from being too lengthy and data-demanding. I strongly believe that short, coherent lab information with plenty of instructive graphics in a book or e-book format is a handier alternative than scattered and fragmented internet videos and texts. Hopefully, this book will prove me right.

June 2026

Thomas Norberg
Uppsala, Sweden

Acknowledgments

Contributions and advice from many friends and colleagues have been essential for the completion of this book project. Special thanks to Helena Grennberg, Elisabet Kallin, Julia Eklund, Lukasz Pilarski, Mate Erdelyi, Jan Streuff, Gunnar Svensson, Daniel Sohn, Ola Åberg, Christian Hedberg, Oscar Verho, Mats Neihoff, Margret Holtz, Karl-Gunnar Noren, and Niklas Alveskog.

I would also like to thank my closest collaborators at Wiley, Jonathan Rose, Anne Brennführer, Agasthidevi Arumugavel, and Priyadarshini Natarajan, for being so helpful and professional.

Finally, I thank the Department of Chemistry at Uppsala University, Sweden, for generously providing me with lab space and infrastructure during the years that I have worked with this book.

Thomas Norberg

1

Introduction

What is organic chemistry? If you have ever taken a lab course in organic chemistry, you know that it is very much about “reactions.” “Running a reaction” means that you mix chemicals so that they can react with each other while you check (“monitor”) what is happening with an analytical method. You then “work up” the mixture, meaning that you take care of any new chemical that has been produced and remove any remaining unwanted chemicals, by-products, solvents, etc. Finally, you check the purity and identity of the new chemical(s) with some analytical method(s), and document your actions and results in a notebook or something similar. Taken together, the “running” and “workup” is called “performing a synthetic step” (Figure 1.1).

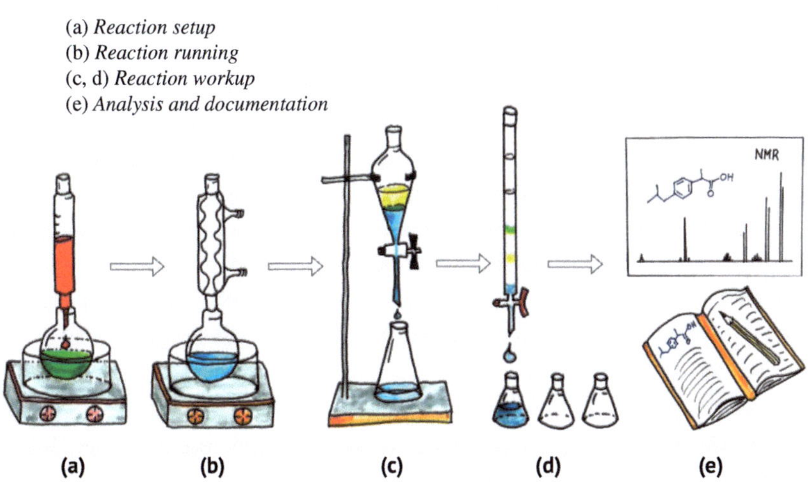

Figure 1.1 Typical workflow when you perform a synthetic step. Drawing courtesy of Elisabet Kallin.

At the molecular level, a reaction is when one or more bonds in a molecule are broken, and new bond(s) are formed, resulting in a new molecule. This could be the molecule you want, but it could also be an intermediate molecule in a whole series of reactions ("synthetic steps") that end with your "target molecule." In the latter case, we are talking about a "multistep organic synthesis." We will return to this "molecule-assembly" activity in a later chapter, but first, some basics about performing a single synthetic step.

There are thousands of known organic reactions where the outcome can be reasonably well predicted when you know the starting chemicals and the conditions; many of them even have names like the Grignard reaction or the Wittig reaction. Every year, new reactions are discovered and described, further expanding the possibilities for organic chemists. Together, the enormous body of known "reactions" and their relationships make up much of the theoretical framework of organic chemistry.

However, when it comes to the laboratory practice ("preparative organic chemistry"), you will find that the equipment and procedures for carrying out the "reactions" are surprisingly few and rather simple, and, believe it or not, some have remained almost unchanged for a long time. Actually, much of the beauty of preparative organic chemistry is that it is possible to assemble, step by step, novel and very complex organic molecules using only a few simple procedures and relatively inexpensive equipment. This is in contrast to the situation today in many other branches of science, such as physics or astronomy, where expensive and complicated equipment are required to make any form of scientific progress. Of course, modern preparative organic chemistry has benefited enormously from the development of sophisticated analytical instruments (such as NMR and MS) that are now indispensable in any synthetic work. This text, however, centers around the basic organic chemistry lab procedures/tools and describes them in detail with emphasis on their simplicity. The aim is to help students and researchers to improve their skills in synthetic organic chemistry.

To put organic chemistry lab techniques in context, it helps to first take a look at how synthetic steps are described in experimental sections of scientific papers. Such texts are written with the assumption that the readers are organic chemists who are familiar with basic lab procedures, so these are simply referred to by "keywords" ("extracted," "flash column chromatography," etc.). A good example is the synthetic step shown in Figure 1.2, described in a paper from 2006 [1]. The experimental section for this step reads as follows (the keywords are highlighted in bold):

> *4-Iodo-7-oxo-6-aza-bicyclo[3.2.1]octane-6-carboxylic Acid tert-Butyl Ester (6).*
>
> Triethylamine (26.2 g, 120.2 mmol), N,N-dimethylaminopyridine (563 mg, 4.6 mmol) and di-tert-butyl dicarbonate (26.2 g, 120.2 mmol) were added sequentially to a solution of iodo lactam 5 (15.2 g, 60.5 mmol) in dichloromethane (100 mL) at 23 °C. The reaction mixture was **stirred** for 4 h and then saturated aqueous ammonium chloride (100 mL) was carefully added. The mixture was diluted with water (100 mL) and **extracted** with dichloromethane (3 × 200 mL). The combined extracts were **washed** with brine (100 mL), **dried** (magnesium sulfate) and **concentrated in vacuo**. The residue was purified by **flash column chromatography** (ethyl acetate:hexanes, 1:4) to afford iodo lactam 6 (20.9 g, 98%) as a colorless solid.

Figure 1.2 The N-acylation synthetic step. *Source*: [1] with permission of American Chemical Society.

Already in this short text, we can identify six "keywords." In the following chapters, we will describe the procedures behind the most common keywords and technical terms in enough detail that an average college student could easily understand them (see also "glossary of terms and keywords" at the end of this book).

When reading the text above, it is also important to note what is NOT mentioned, e.g., the reaction progress must have been monitored by some analytical technique (probably TLC), which is a normal routine. A chapter discussing different ways of "monitoring" reactions is included in this book (TLC, HPLC, MS, and NMR).

Even more importantly, the text does not mention anything about the risks associated with running that particular synthetic step. Lack of such information is common in the literature; the assumption is that whoever tries to carry out the described procedures should be familiar with the risks. This might not always be the case, so a chapter about the safety aspects of practical organic chemistry is also included in this book.

Reference

1 Yeung, Y., S. Hong, and E.J. Corey, A short enantioselective pathway for the synthesis of the anti-influenza neuraminidase inhibitor oseltamivir from 1,3-butadiene and acrylic acid. J Am Chem Soc, 2006. 128: pp. 6310–6311.

2

Reaction Setup and Running

2.1 General Considerations

To set up and "run" a reaction is probably the most fundamental of all organic chemistry activities (Figure 2.1 shows a stirred and cooled "reaction flask"). Typically, it means mixing a solution of some chemical(s) with some other chemical(s) under controlled conditions, and then expecting some "action." Often, you check ("monitor") what is happening over time. When nothing more seems to happen, you consider the reaction finished and proceed with the "workup," where you isolate and purify any new chemical(s) that may have formed. With some reactions, notably those that use excess of very reactive chemicals, you may need to "quench" the reaction before you start the workup. Quenching means that you add something that neutralizes any remaining reactive chemical that would otherwise have caused mishaps during the workup.

Figure 2.1 A reaction flask cooled by a dry ice bath.

The conditions during the actual reaction (temperature, environment, reaction time, etc.) can vary quite a bit depending on the nature of the chemicals involved. In the following sections, we will discuss different variants of reaction setups and running, starting with the simplest. Different methods of reaction monitoring are discussed in Chapter 3. To facilitate reaction monitoring, it is a good habit to withdraw a small sample of the reaction solution immediately before the reaction is initiated. This sample can be used later during the reaction as a reference.

Before we begin, and this applies to all variants of reaction setups that we will discuss below, and also all workup procedures discussed in Chapter 3: Safety first!

Protect yourself and know what you are doing! The most important *personal protection* rule is that you should always wear safety goggles when you are in an organic chemistry lab! It does not take much splashing of a corrosive chemical into the eye to cause irreparable damage, and lost vision can never be replaced. Make it a habit to put on your goggles when you enter the lab.

The most important *general* rule is to do your homework before you start an experiment. Given today's information resources, there is no excuse for not being informed about the dangerous properties of the chemicals and solvents involved. Not only should you be well informed about this beforehand, but you should also take all necessary precautions to minimize risks before the reaction, during the reaction, and during the workup of the reaction. In Chapter 7, you will find the "10 commandments of organic chemistry lab work" plus more detailed information about lab safety issues.

2.2 Reactions at Room Temperature

This is the most common and simplest of all setups. No heating, cooling, or inert atmosphere is required, just a stirred solution of one or more chemicals in a flask to which you add another chemical (in solution or as a neat liquid), often dropwise (Figure 2.2a). A "pressure-equalizing" dropping funnel enables the use of a reaction flask with only one neck. Alternatively, you can perform the addition to a one-necked reaction flask with a syringe needle through a septum, where a venting needle has also been inserted to prevent the build-up of pressure in the flask (important!). For further information about the use of a septum, see Chapter 8. You can,

(a) (b) (c)

Figure 2.2 (a) Room temperature, (b) ice-cooled, and (c) dry ice-cooled.

of course, also use a two- or three-necked flask and an ordinary dropping funnel for the addition. Stirring is usually performed with a magnetic stirrer (Figure 2.2a). At larger scales (> 1 L solvent), reactions might require mechanical stirring with paddles connected to an overhead electric stirrer motor. Read more about stirring in Chapter 8.

2.3 Reactions at Lowered Temperature

Many reactions need to be cooled. Cooling is most often needed to absorb some of the heat evolved, but it can also be done to favor certain reaction products and suppress others. Ice-water is a common cooling medium. The setup is similar to a room-temperature setup, but the reaction flask is immersed in an ice-water bath during addition (Figure 2.2b). If strict temperature control is necessary, a two-necked flask with both a dropping funnel and an inserted thermometer can be employed. With ice-water cooling, the reaction temperature can be kept between +5 and 0 °C. The temperature range can be extended downward to approximately −20 °C by using an ice-water–salt mixture instead of just ice-water.

After the addition is complete and the need to cool the reaction is no longer required, the cooling medium can be removed, and the reaction can warm to room temperature.

Some reactions require temperatures lower than −20 °C to run well. Baths with ethanol–dry ice or acetone–dry ice give temperatures down to −70 °C, and liquid nitrogen baths can go down as low as −196 °C. For convenience, dry ice and nitrogen baths should be well insulated, for example, with a Dewar container or similar (Figure 2.2c).

2.4 Reactions at Elevated Temperature

Some reactions need heating to proceed at a measurable rate. The most common arrangement is to boil the reaction mixture while also stirring it using a magnetic stirrer hotplate (Figure 2.3). The solvent vapors are condensed back into the reaction flask by a "reflux condenser" attached vertically to the top of the reaction flask; the entire process is called "refluxing." The heat source is most often a PEG bath (see Chapter 8) placed on the magnetic stirrer hotplate. The bath temperature is controlled by a bath thermostat, which has a sensor dipped into the bath.

A water bath can be used instead of a PEG bath if the reaction solvent has a boiling point well under 100 °C and the reaction time is short enough that the water does not evaporate significantly.

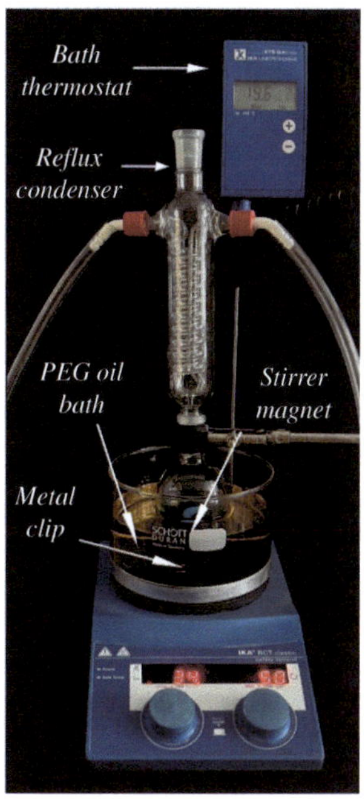

Figure 2.3 Reaction at elevated temperatures.

2.5 Reactions Under Exclusion of Moisture

Some reactions may require strict absence of moisture but can be carried out in the presence of air. Flame-drying of the reaction vessel (see Section 2.6, Figure 2.5a) and the use of dry solvents may not be enough to exclude all moisture, and an internal drying agent is therefore added as an extra precaution. This drying agent should be of a type that does not interfere with the reaction itself. One popular such agent is "molecular sieves" in the form of pellets or powder [1]. Because of its strong water-absorbing properties, this material requires pre-drying shortly before use; otherwise, it may actually add moisture to the reaction. Pre-drying is best done at 230–250 °C under vacuum (several hours), see Figure 2.4.

When molecular sieves are used in reactions, one must always consider that they are also acid-adsorbing, which may or may not affect the reaction outcome.

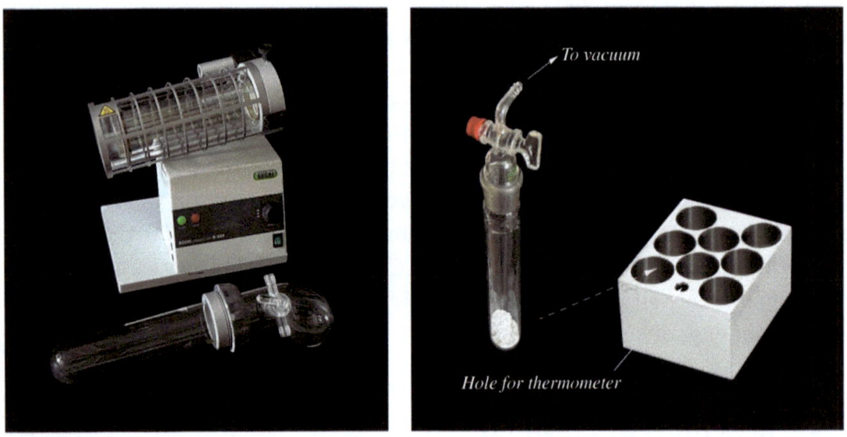

(a) (b)

Figure 2.4 (a) Example of an "oven" capable of drying molecular sieves under vacuum at 230–250 °C. (b) A simpler arrangement for drying molecular sieves under vacuum at 230–250 °C. An aluminum block, with a glass tube and a thermometer inserted into their respective holes, is placed on a hotplate where the temperature can be regulated.

2.6 Reactions Under Exclusion of Moisture and Air

2.6.1 Introduction

Many chemicals are sensitive to both moisture and air, and their handling therefore requires special arrangements to exclude both these elements. This is usually accomplished by running the reactions in pre-dried glassware under dry inert gas. Actually, working with organic synthesis under such conditions has become something of an art in itself, employing everything from simple equipment like syringes and inert-gas balloons to expensive and sophisticated glove boxes. For more comprehensive discussions of this subject, see dedicated treaties [2–4].

In the following, we will exemplify how an air/water-sensitive liquid can be added to a solution of another chemical under exclusion of air and moisture (lab scale, < 1 g). However, first a few words about inert gases: The most popular inert gases are argon and nitrogen. Argon is the best choice, but it is expensive. Argon is *very* unreactive (it is a Group 18 element!), and it is also quite heavy, forming a layer over a liquid surface, excluding lighter gases. The second-best choice is nitrogen. It is much cheaper and is, with a few exceptions, also unreactive. High-pressure argon and nitrogen gases are sold in steel cylinders as "anhydrous" (very little water vapor can contaminate at gas cylinder pressures). Despite this, many chemists connect an extra gas drying device ("drying tower") to their gas cylinder outlet, to ensure complete dryness.

In the following example (Figure 2.5a–k), dry inert gas is delivered from a steel cylinder's reduction valve via a double manifold ("Schlenk line"). This device, connected to both an inert gas cylinder and a vacuum pump, provides both the inert gas and the vacuum needed for the evacuation of air. For a detailed description of a Schlenk line, see Chapter 8. Another alternative for inert gas and vacuum delivery (not shown here) is to use balloons/needles/septa throughout in combination with a gas steel cylinder and a vacuum pump.

2.6.2 Reactions Under Exclusion of Air and Moisture, Step by Step (Figure 2.5a–k)

The first thing to do here is to make sure that the moisture, which is usually adsorbed on the glassware surface, is removed. This is best done by "flaming" the equipment under vacuum. A vacuum oven could also be used, but it is expensive and seldom provides the heat (several hundred °C) required. Figures 2.5a,b describe

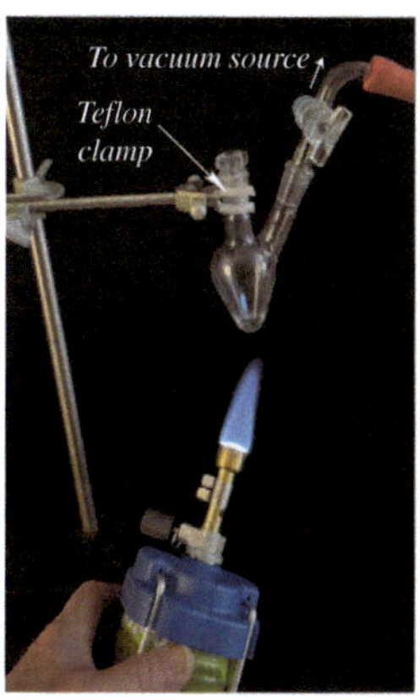

Figure 2.5a Connect the reaction vessel to a vacuum from a Schlenk line and open the tap. Heat it with a burner or heat gun.

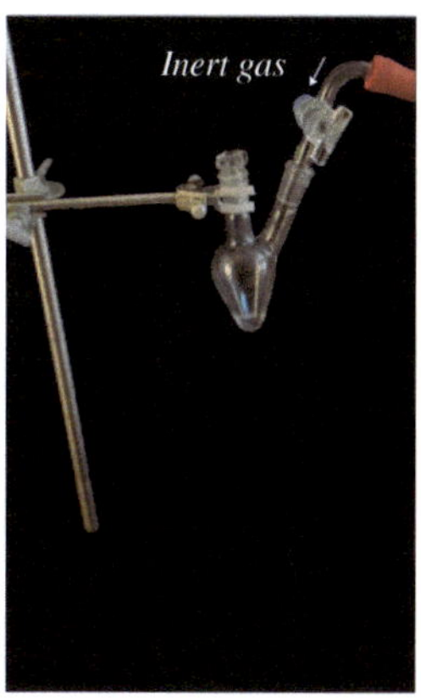

Figure 2.5b When you have heated the glass as uniformly as you can for a few minutes, switch the Schlenk line from vacuum to inert gas.

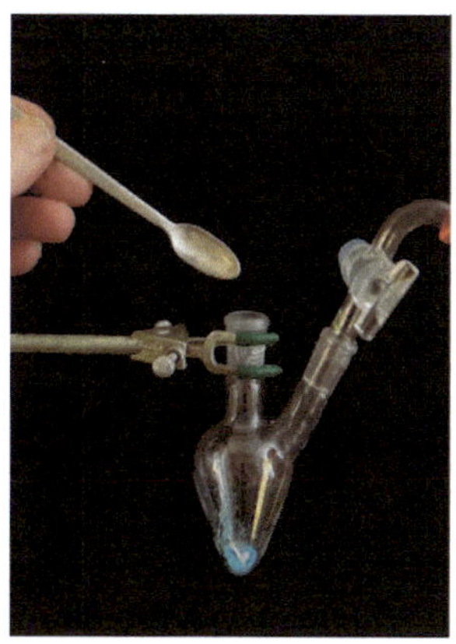

Figure 2.5c Turn off the inert gas supply while you add the first chemical (in this example, it is a solid which is not air- or moisture-sensitive) and a stirrer magnet.

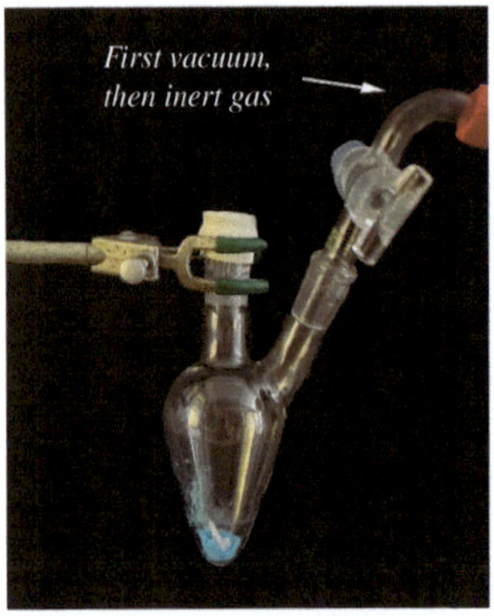

Figure 2.5d Add a septum to the vessel, apply vacuum again, then switch to inert gas. Maintain a slightly overpressurized inert gas blanket all the way until the reaction is regarded as finished and (if needed) quenched.

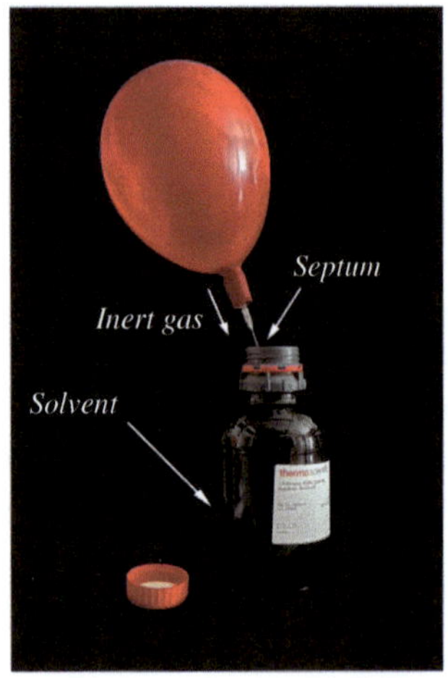

Figure 2.5e The solvent shall now be added. It should be of anhydrous quality and contained in a septum-capped bottle under inert gas. Penetrate the solvent bottle septum with an inert-gas balloon attached to a needle (a Schlenk line inert-gas connection can also be used instead of the balloon).

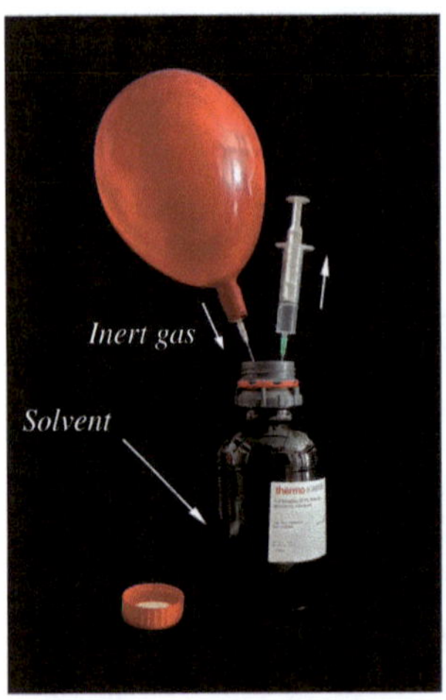

Figure 2.5f Extract the required amount of solvent with a syringe. Before removing the syringe needle completely from the bottle, withdraw it just slightly to bring the needle tip above the liquid level and draw a tiny amount of inert gas into the syringe. Then withdraw the syringe completely from the bottle.

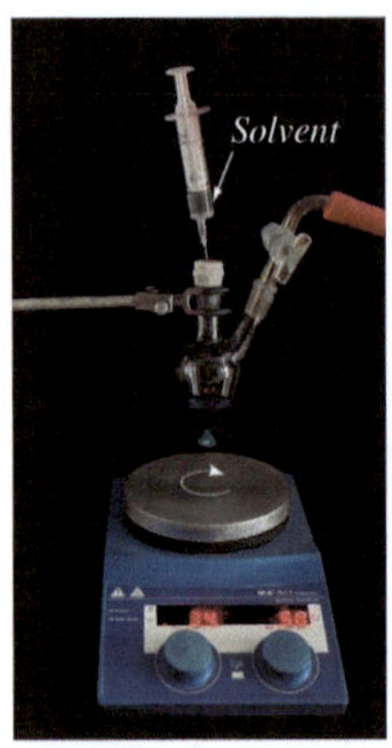

Figure 2.5g Penetrate the reaction vessel septum with the syringe and inject the solvent slowly. Start the stirring.

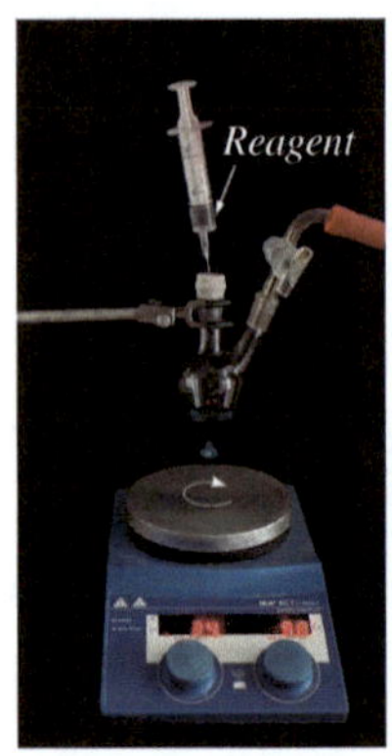

Figure 2.5h The sensitive chemical ("reagent") shall now be added. Use the same syringe needle procedure as before ([e] and [f]) to now extract the required amount of reagent from its bottle. Penetrate the reaction vessel septum, and *slowly* inject the reagent. After finishing the addition, withdraw the needle. Destroy eventual reagent residues in the syringe needle immediately afterward with a quenching reagent.

Figure 2.5i The reaction is now running. For monitoring, you can withdraw *small* aliquots from the reaction vessel with the aid of a thin syringe needle.

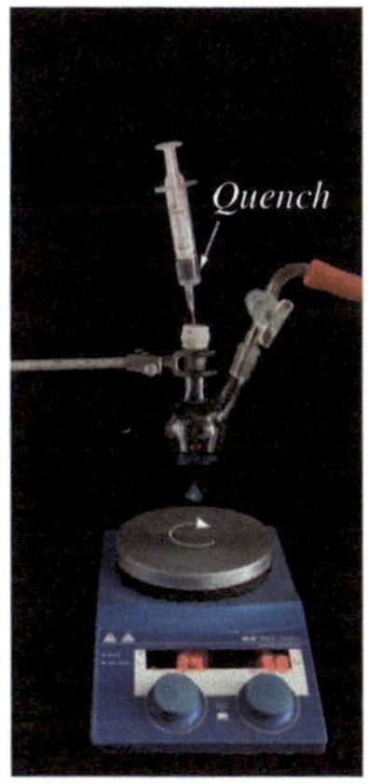

Figure 2.5j When the reaction is regarded as finished, you may need to *slowly* inject a quenching reagent with a syringe needle.

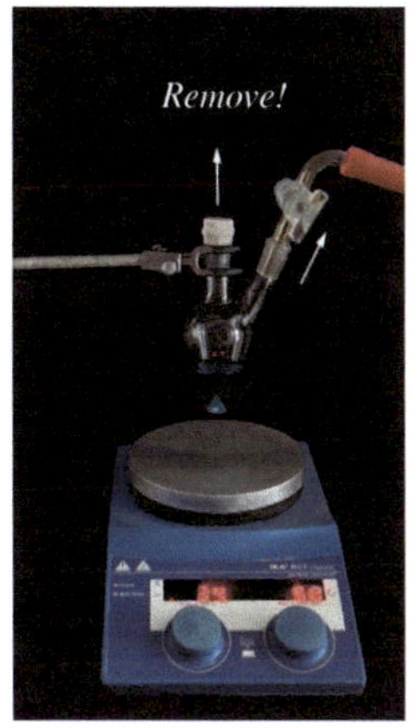

Figure 2.5k After quenching, the reaction mixture should tolerate normal air and can be subjected to normal workup procedures.

the flame-drying process. After flame-drying, the reaction vessel should be used immediately after cooling to room temperature.

If the reaction is to be carried out at room temperature, you can now proceed from Figure 2.5g directly and add the sensitive chemical (Figure 2.5h).

If the reaction needs to be cooled, you must first immerse the reaction flask from Figure 2.5g in a cooling bath (see Section 2.3), allow the mixture to cool, and then proceed (Figure 2.5h).

2.7 Catalytic Hydrogenation Reactions

2.7.1 Introduction

Catalytic hydrogenation is a reaction between hydrogen and a solution of an organic compound (the "substrate") in the presence of a catalyst, which can be insoluble or soluble. Usually, the catalyst is a metal or metal derivative. Catalytic hydrogenation is a *very* common reaction in organic chemistry. It is associated with certain risks since it involves handling of hydrogen, a flammable gas. Since hydrogenations are risky and often require special equipment, the reaction type is here discussed in a separate section. A typical hydrogenation reaction is seen in Figure 2.6.

In the oldest form of catalytic hydrogenation, a solution of the substrate and a solid catalyst is stirred under hydrogen gas at ordinary atmospheric pressure. We will first discuss this type of hydrogenation. Second, we will discuss

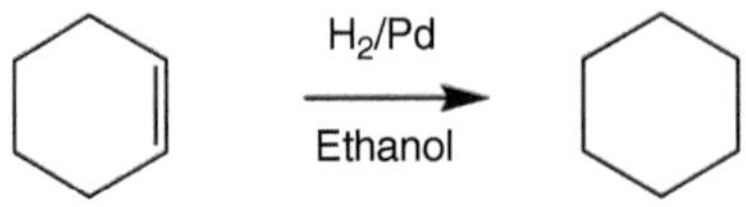

Figure 2.6 Catalytic hydrogenation of cyclohexene to form cyclohexane.

hydrogenation with compressed hydrogen gas and the equipment needed. Third, we will discuss catalytic transfer hydrogenation, which uses a molecule other than hydrogen gas as the hydrogenation reagent. First, some words about risks and safety.

2.7.2 Risk Management During Catalytic Hydrogenations

Mixtures of hydrogen gas and air in a broad range of concentrations are hazardous; only a small spark or flame can result in a violent explosion. Therefore, there must be no air present in a hydrogenation vessel during the entire reaction, and the remaining hydrogen gas must be safely vented afterward. Some catalysts are also fire hazards themselves. When dry, they can catch fire, alone or in contact with organic matter. Raney nickel is an example. Therefore, disposal of hydrogenation catalysts has to be done properly, and most labs have special rules and containers for this. Never let used catalysts become completely dry during workup! Finally, all work should be conducted in a well-ventilated hood, and normal personal protection should be worn (safety goggles and a lab coat). A dry-ice fire extinguisher should also be nearby. A dedicated room for all types of hydrogenations is common and recommended for larger labs.

2.7.3 Atmospheric Pressure Hydrogenation with Hydrogen Gas

For this type of hydrogenation, you can use an ordinary glass reaction vessel with magnetic stirring if air can be removed/excluded and the hydrogen gas provided safely. Here, we describe two common setups: gas delivery with a balloon (Figure 2.7a) or a gas burette (Figure 2.7b).

The balloon delivery is the simplest, but it is controversial for safety reasons (the balloon can burst!). The balloon must be of good quality (see Section 2.7.8). In any case, you should avoid filling balloons with more than 1 L of hydrogen gas! Balloon gas delivery is primarily used for small-scale reactions.

A gas burette like the one shown in Figure 2.7b can hold several liters of gas (the one shown is a 100 mL mini-version), and they can be refilled, if necessary, during a reaction. Burettes can be purchased or constructed from an ordinary measuring cylinder.

In addition to the equipment shown in Figure 2.7a,b, you need hydrogen and nitrogen (e.g., from gas cylinders).

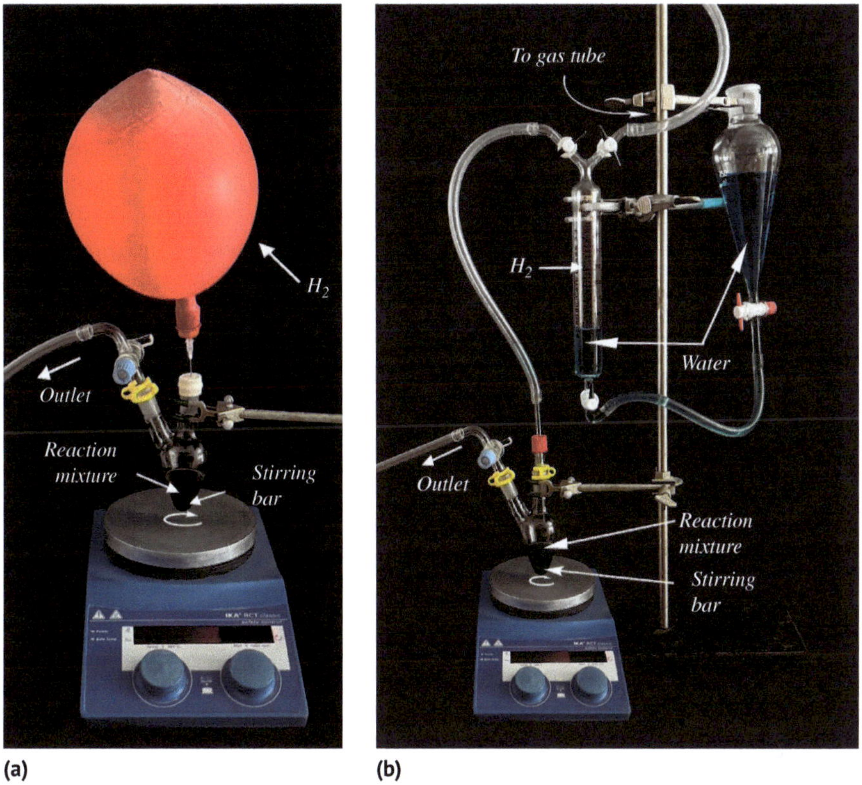

Figure 2.7 Hydrogenation in process with different hydrogen gas deliveries: (a) balloon and (b) gas burette.

2.7.4 Step-by-step Hydrogenation (Balloon)

Please note: do not fill a balloon with more than 1 L (41 mmol) of hydrogen gas at a time. If the calculated consumption of hydrogen gas is higher than 1 L, then refilling the balloon with hydrogen several times can be considered.

1. Equip a two-necked reaction vessel with a magnetic stirring bar and a septum on one of the necks (see Figure 2.7a).
2. Through the second neck, add first the catalyst, then a solution of the substrate. The order of addition is important! A fine dust of dry catalyst can ignite solvent vapors.
3. Attach the outlet tap and clip (see Figure 2.7a) to the open neck and close the tap.
4. Fill a needle-equipped balloon with nitrogen gas, empty it, and refill again several times with the vessel volume of nitrogen gas.
5. Punctuate the septum with the balloon needle, but keep the needle tip well above the liquid!

6. Slowly open the outlet tap and let the nitrogen gas flush the reaction flask until the balloon is almost empty. The gas flow should reach the liquid surface, which is indicated by surface movements.
7. Close the outlet tap and withdraw the balloon needle.
8. Empty the balloon of nitrogen gas completely, refill it with a small amount of hydrogen gas, empty again (in a safe place!), and refill again with hydrogen (no more than 1 L!).
9. Check that the outlet tap is closed, then carefully punctuate the septum with the balloon needle. Keep the needle tip well above the liquid! Slowly open the outlet tap and let a few vessel volumes of hydrogen gas flush the reaction flask, then close the outlet tap.
10. Start the stirring, the rate should be rather fast, and let the reaction run while checking if the hydrogen balloon shrinks. If necessary, the balloon needle can be withdrawn, the balloon refilled with more hydrogen gas, and the balloon needle inserted again.
11. To monitor the reaction, punctuate the septum with a thin syringe needle and withdraw a tiny aliquot of the reaction solution.
12. When the reaction is finished, remove the hydrogen balloon needle, empty it safely, refill it with nitrogen, empty, and refill again for several times with the reaction vessel volume of nitrogen.
13. Punctuate the septum with the nitrogen balloon needle (keep the needle tip well above the liquid), slowly open the outlet tap, and let the nitrogen gas flush the reaction flask until the balloon is almost empty. Stop the stirring.
14. It should now be safe to open the flask and withdraw the reaction mixture. Filter off the catalyst with a fine-pore glass frit filter or a Celite filter pad (see Chapter 4, Section 4.3.3), and wash with solvent. Do not let the catalyst go completely dry during the filtration process! Collect the combined filtrate, and discard the catalyst in a safe way (see Section 2.7.2).

2.7.5 Step-by-step Hydrogenation (Gas Burette)

The procedure is similar to the one above, but a gas burette (Figure 2.7b) instead of a balloon delivers the gas. *Important*: the gas left in the burette after use should be nitrogen!

1. Perform the same steps as 1–3 above, but instead of a balloon, attach the tube from a gas burette, for example, via a ground glass adapter penetrated by a Pasteur pipette (Figure 2.7b). Do not forget the clip! Make sure that the pipette tip is well above the surface of the liquid, and that the two burette upper taps are closed.
2. Empty the gas burette completely by opening the burette's upper left tap and the outlet tap. If necessary, raise the separatory funnel. Close the upper left tap again, and adjust the separatory funnel so that the two water surfaces are level. Close the outlet tap.
3. Connect the burette upper right tap tubing to a nitrogen gas source and fill the burette with nitrogen gas by opening the upper right tap and the nitrogen source tap. When the burette is full, close the nitrogen source tap and the upper right tap. Flush the reaction vessel with the burette nitrogen gas by opening

first the burette upper left tap and then, with care, the reaction vessel outlet tap. Then close the two taps again.

4. Repeat the nitrogen flushing procedure (step 3) once again to make sure that all air in the reaction vessel is displaced.
5. Shift the gas source to hydrogen gas (make sure that no air enters the tubing or burette during shifting). Fill the burette with hydrogen gas by opening the burette upper right tap and the hydrogen source tap. When the burette is full, close the hydrogen source tap and the upper right tap. Flush the reaction vessel with the burette hydrogen gas by opening first the burette upper left tap and then, with care, the reaction vessel outlet tap. Then close the two taps again.
6. Refill the burette with hydrogen gas by repeating step 5. Open the burette upper left tap and start the reaction vessel stirring (it should be pretty fast). Let the reaction run while checking the liquid level in the burette. If necessary, refill the burette with more hydrogen gas (step 5).
7. The reaction can be monitored by following the consumption of hydrogen gas. When the theoretical amount has been consumed, and the burette liquid level has stabilized, shift the gas source to nitrogen, fill the burette, and flush the reaction vessel according to step 3.
8. It should now be safe to open the reaction vessel and withdraw the reaction mixture. Filter off the catalyst with a fine-pore glass frit filter or a Celite filter pad (see Chapter 4, section 4.3.3), and wash with solvent. Do not let the catalyst go completely dry during the filtration process! Collect the combined filtrate, and discard the catalyst in a safe way (see Section 2.7.2).

2.7.6 Elevated Pressure Hydrogenation with Hydrogen Gas

In many labs, elevated pressure hydrogenations with a "Parr" shaker apparatus (Figure 2.8a) is the standard procedure, regardless of substrate and reactivity.

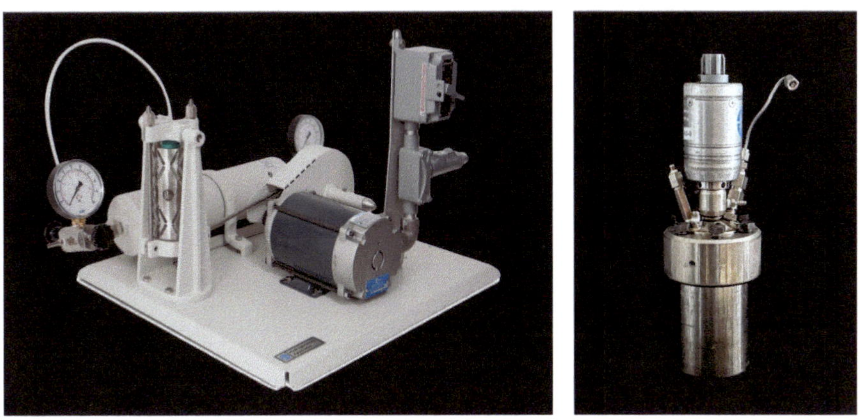

(a) (b)

Figure 2.8 (a) Shaker hydrogenator. Courtesy of Parr Instrument Company. (b) Stirred steel cylinder hydrogenator.

The reaction vessel is usually a 250 mL or 500 mL thick-walled Pyrex glass bottle that can be shaken during the reaction (it can also be heated). Smaller reaction vessels can be fixed inside the bottle if necessary. The apparatus has its own hydrogen gas reservoir, which can be refilled from a standard hydrogen gas cylinder. The maximum working pressure is usually 5 bar.

Depending on the apparatus model, the operation procedures differ slightly. They usually include air removal by evacuation, vessel filling with hydrogen gas, reaction/shaking, and final removal of hydrogen gas by evacuation.

Another pressure solution is a stainless-steel cylinder with a stirring device (Figure 2.8b). The maximum working pressure here can be very high, up to ordinary hydrogen gas cylinder pressure (300–350 bar).

Advantages of pressurized hydrogenations are that the reactions proceed faster and start easier than atmospheric pressure hydrogenations. The main disadvantage is that reaction monitoring is more difficult. Extreme pressures can also give unwanted "over-hydrogenations" of aromatic rings.

2.7.7 Catalytic Transfer Hydrogenation

Some substrates can be catalytically hydrogenated by a hydrogen donor other than hydrogen gas. Examples of such donors are formic acid (or its salts) and isopropanol. Reactions are carried out simply by stirring a solution of the substrate in the presence of the hydrogen donor and a catalyst. Temperature and concentrations can vary depending on substrate and catalyst; room temperature is often sufficient. Sometimes the donor also serves as a solvent (as in the example in Figure 2.9).

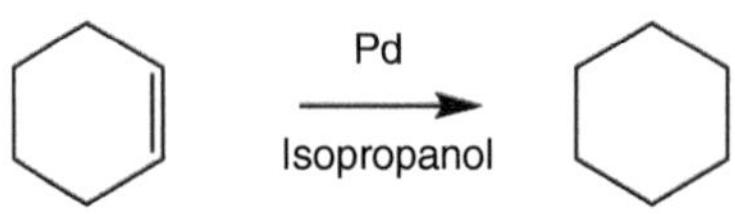

Figure 2.9 Catalytic transfer hydrogenation of cyclohexene to form cyclohexane.

Catalytic transfer hydrogenation is, if applicable, the simplest and probably safest of the hydrogenation procedures. However, the risks with catalyst handling are similar to those during gas hydrogenations. For more details about catalytic transfer hydrogenations, see reviews [5–7].

2.7.8 Tips and Tricks

Balloons: the combination of balloon, needle, and septum for handling smaller amounts of hydrogen gas comes with a risk: balloon bursting. You can minimize this risk by using quality balloons fresh from the package and never using them twice. Toy balloons become unreliable after only a few days of lab use. Double balloons can, with some effort, be produced by squeezing one balloon inside another, giving an extra degree of safety against bursting.

Needle-equipped balloons: see Figure 2.10 for how to connect a balloon to a syringe needle. A large rubber stopper can serve as an "anchor" and flow-stopper for the needle–balloon combination.

Monitoring by observing the burette gas consumption: some hydrogenation substrates (e.g., those with *N*-carbobenzoxy groups) release as much CO_2 as they consume hydrogen, which means that "volumetric" monitoring of a reaction with the gas burette is not possible.

Catalyst filtering: some metal catalysts are notoriously difficult to filter off with ordinary glass-frit filters or Celite pads; the finest particles always seep through. A final filtering through a PTFE filter disk (Figure 2.11) usually solves the problem.

Sluggish reactions: sometimes a hydrogenation refuses to go to completion or even to start. Especially, atmospheric pressure hydrogenations can be troublesome, and results can vary between catalyst batches. Fortunately, there are some tricks you can try here: (i) increase the pressure, i.e., switch to a Parr shaker or a steel cylinder. This is however a major change and may in fact not be necessary. Instead, you could first try to (ii) increase the temperature, or (iii) if the reactant and product can tolerate it, add some acid, e.g., acetic acid. (iv) Change to another batch of the catalyst, activities can vary significantly between batches. (v) Change the solvent. Many chemists prefer neat acetic acid.

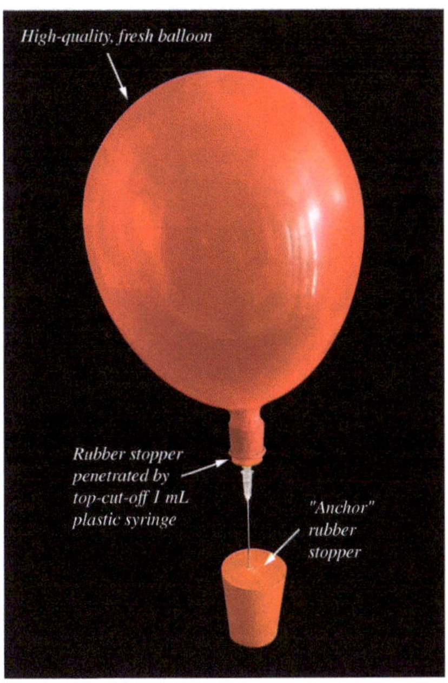

Figure 2.10 A needle-equipped balloon.

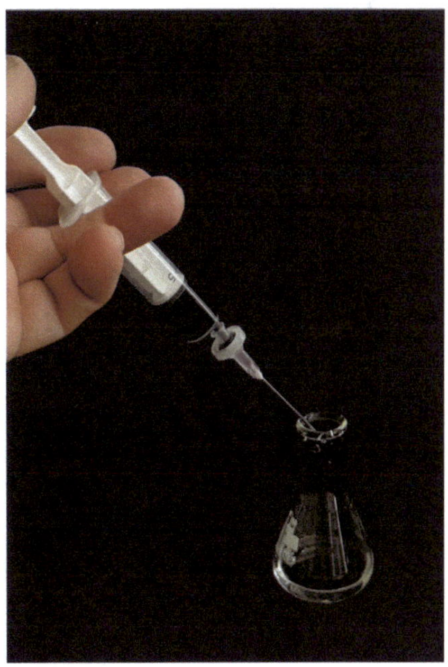

Figure 2.11 Filtering through a PTFE filter disk.

2.8 Solid-phase Reactions

2.8.1 Introduction

The reactions that have been discussed so far are all examples of "solution-phase" reactions, where chemicals are brought together to react in *solution* (even though the catalyst could sometimes be a solid). However, in some situations, it is actually better to have one of the reacting chemicals "anchored" to the surface of an *insoluble* material (the "solid phase") and then mix with a solution of the chemical(s) that it is supposed to react with. The idea is that the reaction product will still be anchored to the solid phase after the reaction. Such reaction mixtures are *very* easy to work up by simply filtering, leaving the solid phase with its anchored product on the filter! The material can then, if desired, be further transformed in a second reaction, followed by another filtration, etc. After a series of consecutive "solid-phase" reactions, the anchor to the solid phase can be cleaved by a chemical reaction, and the solid phase removed by filtration. The cleaved-off product is then in the filtrate and can be purified by, for example, chromatography.

The solid-phase synthesis concept was developed by Bruce Merrifield in the 1950s and 1960s. He convincingly demonstrated that linear polymers like peptides could be rapidly prepared in high yield by stepwise addition of monomer units to a solid-phase-anchored starting unit. The procedures could be automated, and solid-phase synthesis procedures and machines are nowadays routinely used for rapid chemical synthesis of peptides with up to ~50 amino acids in length [8–10] or nucleic acids of a similar size [11]. Merrifield was awarded the Nobel Prize in Chemistry in 1984 for his discoveries.

A solid-phase synthesis machine is a major investment. When a laboratory carries out, for example, solid-phase synthesis of peptides only occasionally, a manual procedure is often preferred, since it is much cheaper compared to a machine procedure, and requires only *very* simple equipment. Figure 2.12a shows the type of equipment that can be used to produce peptides on a 50–100 μmol scale. Figure 2.12b–d show the major operations needed to carry out one synthetic step: reactant solution addition, agitation during the reaction, and filtering off the reactant solution after the reaction has finished. The solid phase, still in the reaction vessel, is ready for the next synthetic step after a few washings/filtrations with clean solvent.

2.8.2 Tips and Tricks

2.8.2.1 Reaction Vessels

A reaction vessel for manual solid-phase synthesis must not look exactly like the one shown; there are many other alternatives. It should contain an upper and a lower opening and a bottom filter disk; other than that, the vessel design and material can be varied quite a bit. It is quite common to use a glass vessel with a bottom glass frit and an open–close tap connected to the lower opening, as shown in Chapter 3, Figure 3.17.

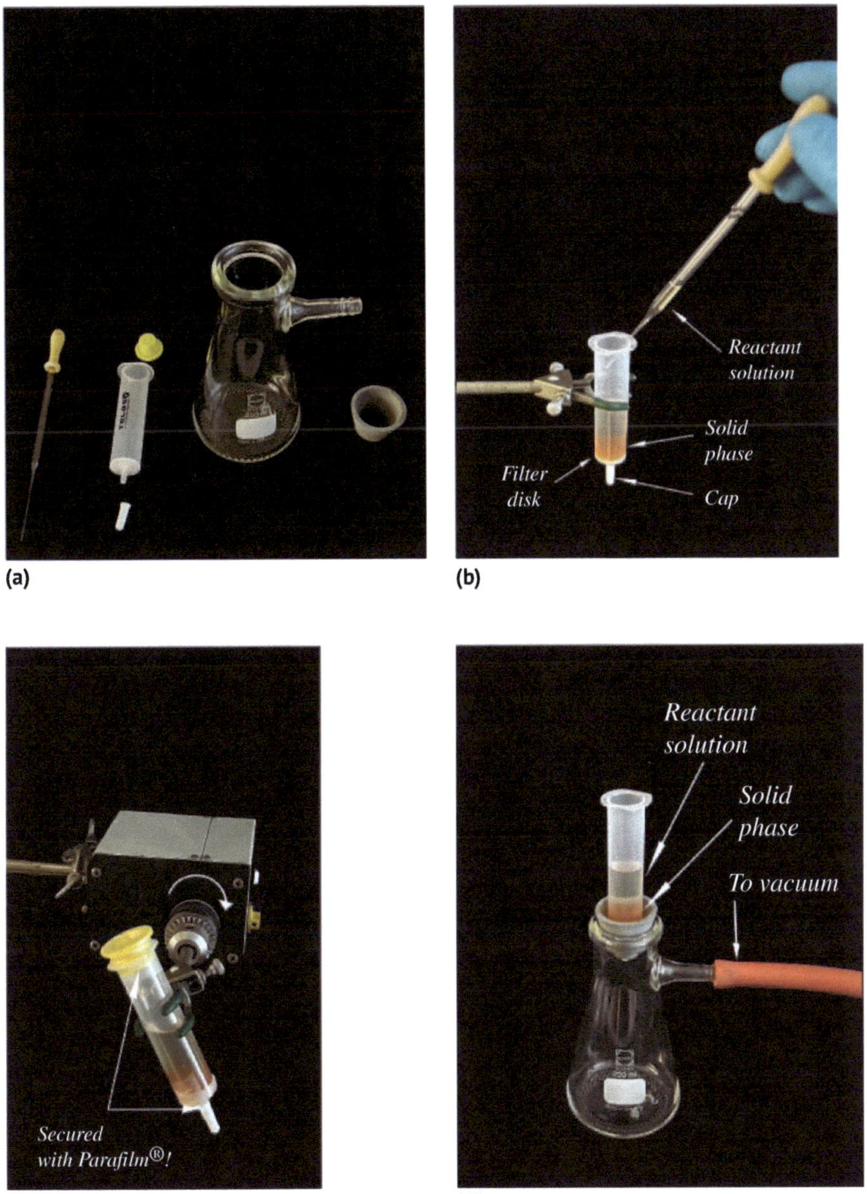

Figure 2.12 (a) Equipment needed for manual solid-phase synthesis, from left to right: pipette, plastic reaction vessel with upper and lower caps and a bottom filter disk, filtration flask, and rubber collar. (b) Addition of reactant solution. (c) Slow agitation during the reaction. (d) Filtering off the reactant solution and washing.

2.8.2.2 Agitation

Solid-phase reactions, like solution-phase reactions, need to be agitated during the reaction for the best results. However, stirring with a magnet or a propeller inside the reaction vessel is not recommended, since it can cause grinding of the solid-phase particles, making them more difficult to filter. Instead, the entire (carefully closed!) reaction vessel can be agitated, for example, by being clamped to the drive shaft of a horizontally mounted stirring motor and rotated at very low speed (Figure 2.12c). Another way is to put the closed reaction vessel on a tilting or rotating laboratory shaker.

2.8.2.3 Chemicals and Instrumentation

Most of the specialty chemicals, solvents, and accessories needed for solid-phase synthesis of peptides or nucleic acids are available from vendors. Different solid-phase materials are offered with or without the first monomer anchored, and needless to say, all the monomers, coupling agents, and solvents that you would ever need are catalogued. Therefore, comparatively little time and effort have to be spent on preparing starting materials like protected amino acids or nucleosides. Repetitive additions, filtrations, and washings are also pretty much routine regardless of whether the synthesis is carried out manually or with a machine. The biggest issue is often the purification of the final product. Each solid-phase reaction can be driven to very near completion by using a large excess of the soluble reactant, but by-products are still formed and accumulate, especially when the synthesis involves many cycles. A preparative HPLC purification of the final product is therefore often needed.

2.8.2.4 Monitoring

A solid-phase reaction cannot, like a solution-phase reaction, be directly monitored by TLC or LC–MS, since both reactant and product are anchored to a solid phase and not present in solution. Other monitoring methods are therefore used. Read more about monitoring of solid-phase reactions in Chapter 3.

References

1 Yasukawa, H., et al., Effect of activation methods of molecular sieves on ketimine synthesis. J Org Chem, 2022. 87(21): pp. 13750–13756.
2 Borys, A.M., An illustrated guide to Schlenk line techniques. Organometallics, 2023. 42: pp. 182–196.
3 Leonard, J., B. Lygo, and G. Procter, *Advanced Practical Organic Chemistry*, 3rd Ed. 2013, Boca Raton, FL: CRC Press. 356 pages.
4 Shriver, D.F. and M.A. Drezdzon, *The Manipulation of Air-Sensitive Compounds*, 2nd Ed. 1986, New York: John Wiley & Sons.
5 Brieger, G. and T.J. Nestrick, Catalytic transfer hydrogenation. Chem Rev, 1974. 74(5): pp. 567–580.

6 Wang, D. and D. Astruc, The golden age of transfer hydrogenation. Chem Rev, 2015. 115(13): pp. 6621–6686.

7 Levy-Ontman, O., et al., Acceleration of transfer-hydrogenation of cyclohexene with palladium catalysts in the presence of polysaccharides. Org Commun, 2020. 13: pp. 138–145.

8 Toy, P.E. and Y.E. Lam, *Solid-Phase Organic Synthesis: Concepts, Strategies, and Applications*. 2012, Weinheim: Wiley-VCH. 560 pages.

9 Sewald, N. and H.-D. Jakubke, *Peptides: Chemistry and Biology*. 2002, Weinheim: Wiley-VCH.

10 Amblard, M., et al., Fundamentals of modern peptide synthesis. In J. Howl, editor, *Peptide Synthesis and Applications*, volume Methods in Molecular Biology, Chapter 1, pages 3–24. Humana Press, 2005.

11 Reese, C.B., Oligo- and poly-nucleotides: 50 years of chemical synthesis. Org Biomol Chem, 2005. 3(21): pp. 3851–3868.

3

Reaction Monitoring

3.1 General Considerations

A common mistake by a beginner organic chemist is to run a synthetic reaction exactly as it is described in the literature "recipe," without following (monitoring) what is actually happening during the reaction. Monitoring may appear unnecessary and time-consuming, especially if there is no mention of it in the literature description. However, the reaction time and outcome are very much dependent on factors such as temperature, moisture, batch size, purity of reagents, particle sizes, and light conditions, just to mention a few. The literature descriptions seldom specify all of these factors, and seldom mention anything about monitoring, despite the fact that it is an important part of the exercise. For consistent results and a deeper understanding, organic chemists should always monitor their reactions by some suitable analytical method.

What methods are suitable for reaction monitoring? They should be fast enough not to delay the execution of a reaction, and yet be powerful enough to provide information about reactants, products, and by-products. They should also use equipment that is available for everyday use. Several methods fulfill these requirements.

Thin-layer chromatography (TLC, Figure 3.1a) has long been the favorite monitoring method in organic chemistry, sometimes in combination with offline mass spectrometry. It is simple, fast, and surprisingly powerful. However, chemists now often use LC–MS (liquid chromatography–mass spectrometry, Figure 3.1b), which is also fast and *very* powerful. It not only separates reaction components, but also provides their molecular weight!

Another powerful monitoring method is NMR spectroscopy. However, it is seldom used for routine monitoring of synthetic reactions. Spectra of reaction mixtures are dominated by solvent signals, and the reactant and product signals can be difficult to single out. Also, everyday use of an NMR spectrometer for monitoring of reactions can be problematic because each spectrometer usually serves many users. On the other hand, NMR spectroscopy is an excellent tool for studies of reaction *mechanisms*, where the focus is more on rates and intermediates and less on speed and simplicity, which are critical for monitoring.

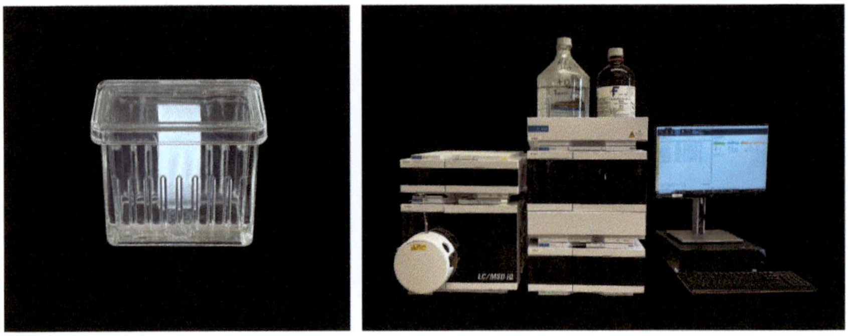

(a) (b)

Figure 3.1 Two common analytical methods for reaction monitoring: (a) TLC and (b) LC–MS.

In conclusion, TLC and LC–MS are today the most commonly used analytical methods for routine reaction monitoring. TLC was the standard for many decades, but LC–MS has grown in importance, especially in labs that have the resources needed to provide reliable everyday access to an instrument.

A special case is monitoring of solid-phase reactions: they require other methods than solution-phase reactions, since the most important reactant in solid-phase reactions is not in solution during the reaction. The last section of this chapter will discuss this issue.

3.2 Thin-layer Chromatography

3.2.1 Introduction

Of all the chromatographic analysis techniques used today, TLC is the simplest and cheapest. It is sensitive, fast, and surprisingly powerful, all of which make it an excellent tool for monitoring the progress of an organic reaction. However, concerning TLC, one can say that, as with chess, it "takes a minute to learn and a lifetime to master." The basic operations can be demonstrated in a student lab and carried out with simple equipment in less than 30 minutes from start to finish, but mastering TLC to its full extent takes years of frequent use. Hopefully, this text can help the reader to shorten the learning time by providing a clear description of the basic operations ("running" and "visualizing") plus some useful tips/tricks. Many articles and books on TLC are available, with the most impressive being "Thin-Layer Chromatography" by Egon Stahl, one of the pioneers of TLC. The book (>900 pages!) is available in both German [1] and English [2] versions. There are also many other books that treat the subject [3, 4].

TLC uses a plate made of glass, aluminum, or plastic covered with a thin (0.1–0.3 mm) layer of a porous material, usually silica gel (Figure 3.2a). This porous material constitutes the "stationary phase." Dilute solutions of the samples to be analyzed are spotted at the bottom end of the plate, and the sample solvent is left to evaporate. The plate is then put in a chamber for "elution." The chamber has a shallow layer of solvent in the bottom. Because of the porous structure of the plate's thin layer, the solvent ("mobile phase") starts ascending in the plate layer by capillary forces, bringing with it the compounds in the applied spot. The trick here is that compounds travel at different speeds

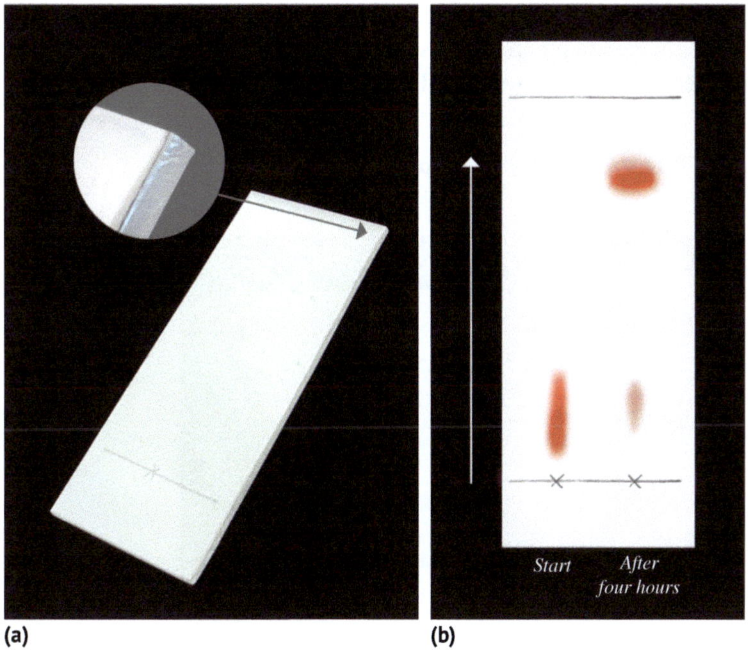

Figure 3.2 (a) Typical TLC plate with an enlargement showing the thin layer on top of the glass. Many types of plates (glass, aluminum sheet, various coatings, etc.) are nowadays commercially available. (b) TLC of a typical reaction mixture (here an amine-to-amide conversion) showing the starting composition (left lane) and the composition after four hours (right lane). The spots are reddish from staining with ninhydrin after elution. The arrow indicates the elution direction.

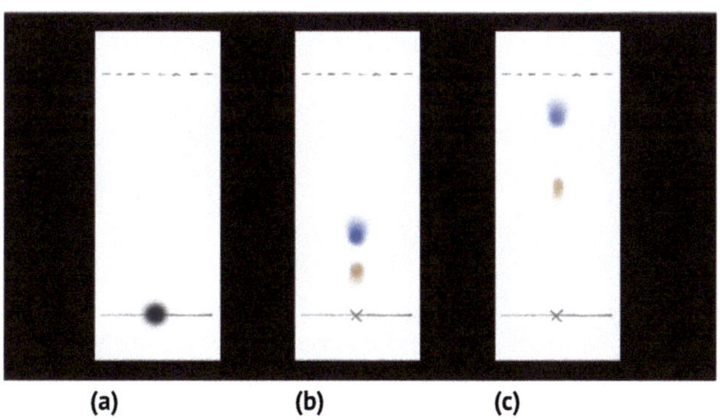

Figure 3.3 (a–c) A TLC plate seen at different timepoints. The sample contains a mixture of two pigments (bluish and yellowish). They adhere differently to the stationary phase, and therefore move upward at different speeds. The entire process takes 5–20 minutes, depending on the length of the plate and the properties of the elution solvent.

up the plate because of their different affinities for the stationary phase. Therefore, they will soon appear as individual "spots" at various distances from the starting point (Figure 3.2b), the distance traveled by a spot divided by the distance traveled by the solvent front is called the R_f value. If the compounds are colored, the elution process can be seen with the naked eye, as in Figure 3.3a–c. In the final picture (Figure 3.3c), the two main components of the sample are nicely separated and clearly visible. However, if the compounds in the sample are colorless, they need to be "visualized" after the elution process. There are several ways to do this, see Sections 3.2.3.6–3.2.3.8.

3.2.2 How to Run a TLC Analysis, Step by Step (Figure 3.4a–l)

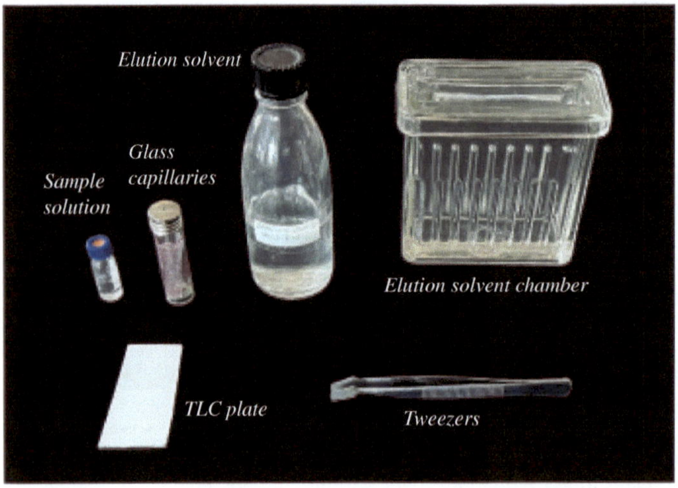

Figure 3.4a Equipment that you need for running a TLC analysis.

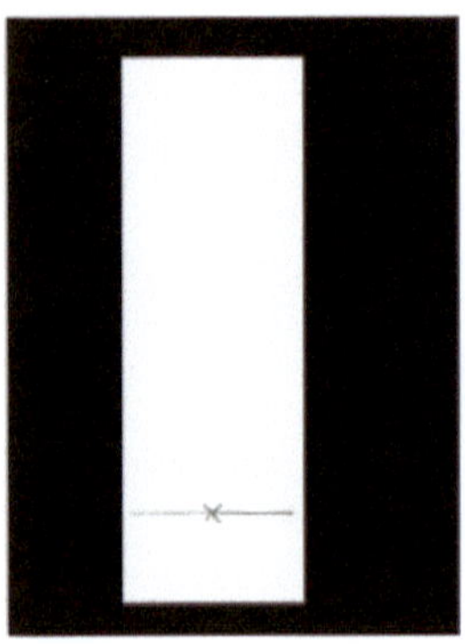

Figure 3.4b Make a pencil line approximately 10–15 mm above the lower end of a suitable TLC plate, then mark the position(s) where you will apply the sample solution.

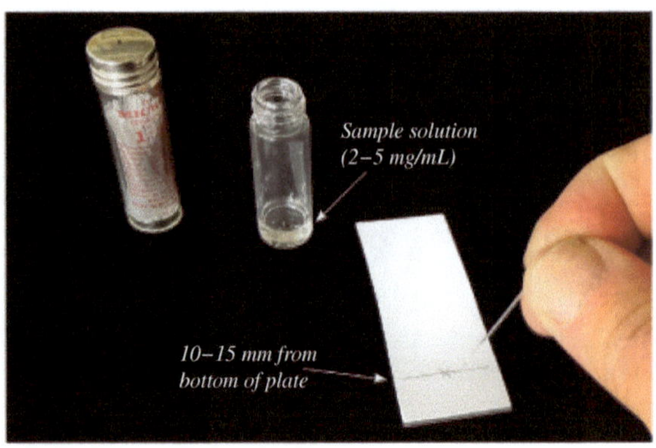

Figure 3.4c Spot a tiny amount (1–5 μL) of the sample solution(s) on the marked position(s) with a glass capillary, and let the solvent evaporate. Note that volatile organic solvents evaporate quickly at room temperature, but solvents like water or DMF may need a vacuum chamber for a minute or two.

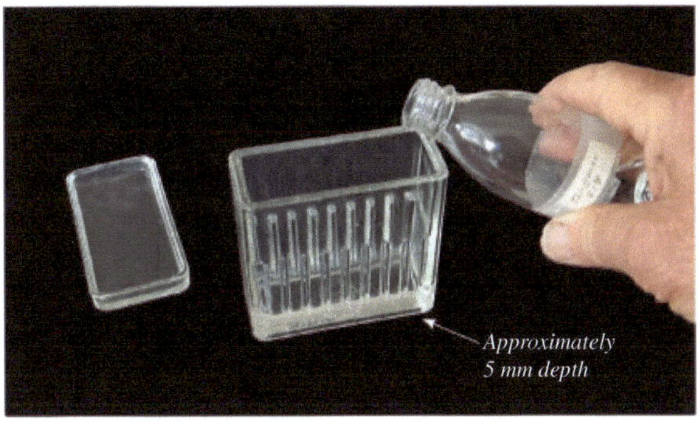

Figure 3.4d Add the elution solvent to the elution chamber, it should form a shallow layer (5 mm deep, not more) at the bottom.

Figure 3.4e Carefully insert the TLC plate into the elution chamber with the sample spots down and close the lid loosely. Caution: the sample spots must *not* dip down under the solvent surface level at the bottom of the chamber!

Figure 3.4f Let the plate develop. When the solvent front has traveled 80–90% up the plate length, the elution process can be considered finished.

Figure 3.4g Take out the plate from the chamber and quickly mark the solvent front with a pencil or similar.

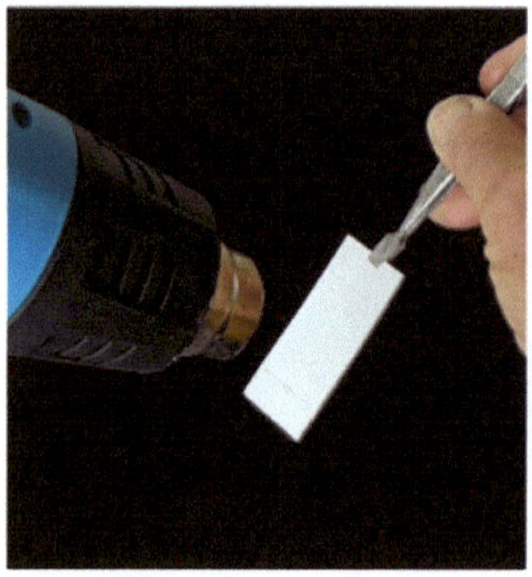

Figure 3.4h Dry the plate with a heat gun or similar. Make sure you do this in a hood or other well-ventilated place. Colored compounds are already visible, but with colorless compounds you have to *visualize*.

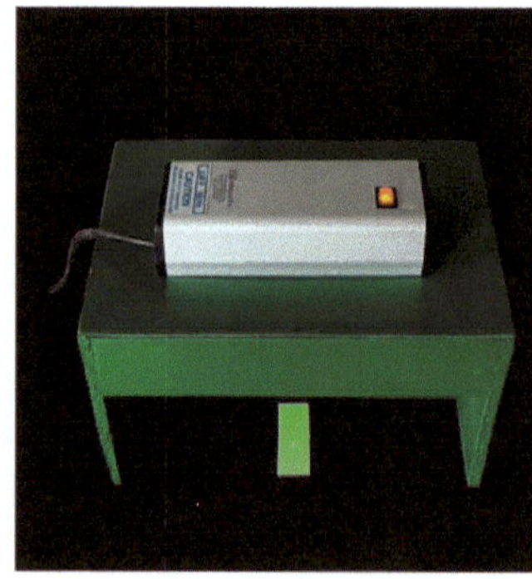

Figure 3.4i Visualizing: inspect the plate under a UV lamp. Any UV-absorbing compounds (at 254 nm) will appear as dark spots. Mark with a pencil or take a photo. Caution: wear safety goggles when working with UV light!

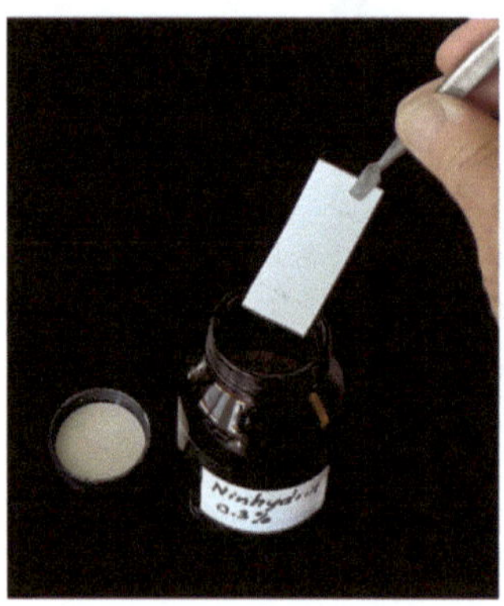

Figure 3.4j All compounds may not absorb UV light, then *visualizing by staining* is an option. Briefly preheat the plate with a heat gun, then dip it quickly in a staining solution (here ninhydrin).

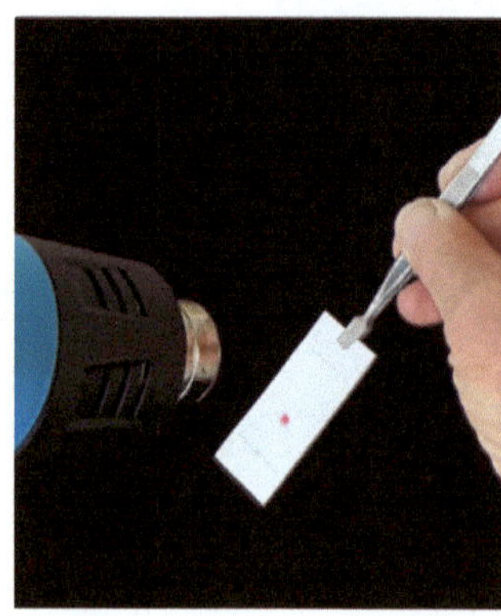

Figure 3.4k Heat with a heat gun until spot(s) appear. There are many different staining solutions, see Section 3.2.3.

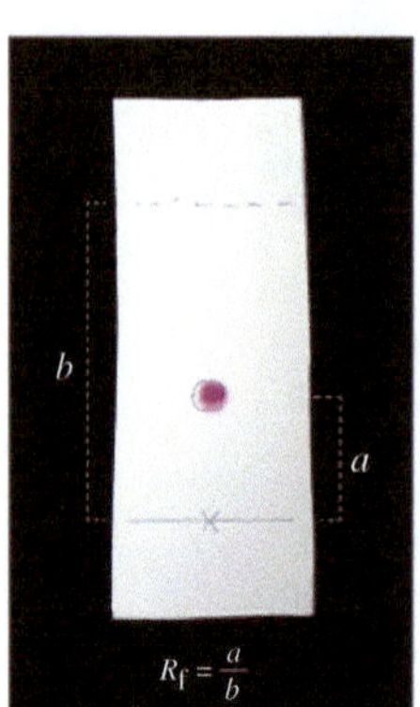

Figure 3.4l The R_f value indicates how fast a compound has traveled up the plate. Here, the sample contained only one compound. In most cases, the sample contains several compounds, each giving a spot with an R_f value.

3.2.3 TLC Tips/Tricks and Other Useful Information

This subsection contains more detailed information about the various steps in a TLC analysis.

3.2.3.1 The TLC Plates

When TLC first became popular in the 1950s, the plates were usually based on silica gel coated onto glass. They were prepared in the lab by dispensing a thin "soup" of silica gel plus additives onto glass plates with the aid of a simple tool, and then the plates were air-dried or "baked" in an oven. The procedure was not particularly difficult to carry out, but when good-quality TLC plates became commercially available in the 1960s, the local lab production of plates was largely abandoned. A detailed description of plate preparation procedures can be found in Stahl's books [1, 2]. It could be useful to know how to prepare your own plates if you want to test new TLC materials and additives.

Nowadays, a large variety of TLC plate types and sizes are commercially available. For the analysis of organic compounds, the silica/glass or silica/aluminum sheet plates are the most popular, usually with a silica layer thickness of 0.10 or 0.25 mm containing a fluorescent additive that enables visualization of UV-absorbing compounds. Although plates in small, ready-to-use sizes are available, the 20 × 20 cm plates that you cut yourself to a desired size are recommended, both for budget and flexibility reasons. One 20 × 20 cm plate is typically cut into 15–30 smaller plates. Aluminum plates are simply cut with a pair of scissors, with due care taken not to damage the thin layer by bending the aluminum sheet. Glass plates, however, have to be cut with a glass cutter. It takes some practice to learn how to use this tool properly, but once you have learned it, you will appreciate the glass plates' better chemical and mechanical stability compared to the aluminum sheet plates.

3.2.3.2 Application of the Sample Solution

Dissolve, if possible, the sample in a highly volatile application solvent (e.g., acetone, chloroform, or dichloromethane). The concentration of the sample should be high enough for the spots to be detected but low enough to avoid spot distortion caused by sample "overloading." In practice, 1–10 mg/mL (w/v) solutions are used for 0.25 mm TLC plates. In Figure 3.5, the effects of different sample concentrations are illustrated.

The application solvent should be allowed to evaporate completely before the plate is put in the elution chamber. Less volatile solvents such as water or DMF require exposure to vacuum (0.1–1 mbar) to evaporate completely at room temperature. If the application solvent remains on the plate, the spot shapes after the elution can really be weird, an example is shown in Figure 3.6.

Finally, a note about analyzing reaction mixtures by TLC: if the TLC spots look bad or distorted, it could be because of excessive concentrations, high-boiling reaction solvents, or interfering reagents and by-products. Then some processing of sampled reaction mixtures might be needed before application. It could be dilution, vacuum

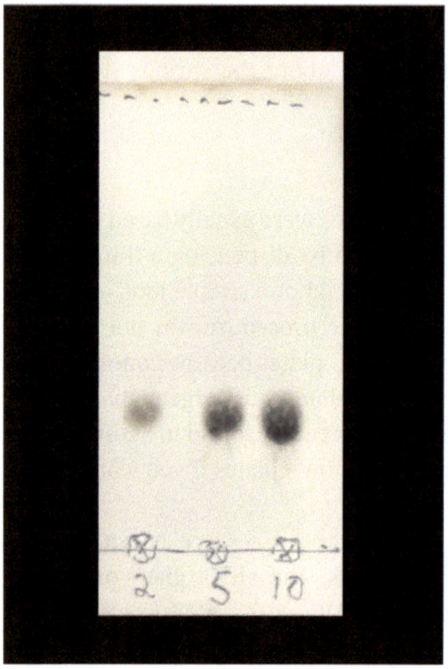

Figure 3.5 An example where the sample concentrations have been varied: lactose 2–10 mg/mL, visualization: sulfuric acid. Elution solvent: ethyl acetate–methanol–acetic acid–water (10:3:3:2). The 10 mg/mL spot shows slight signs of "overloading" (i.e., distorted spot shape).

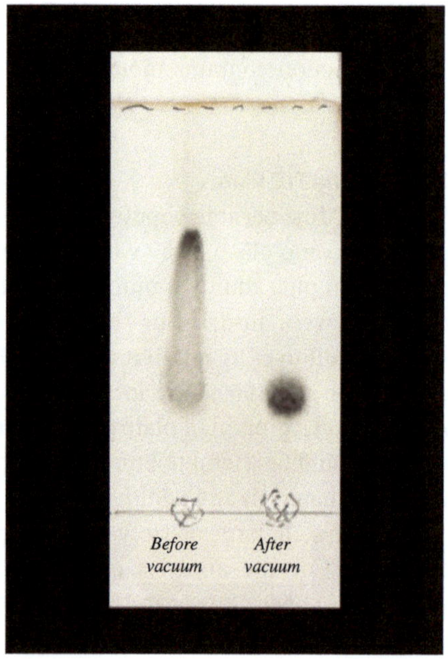

Figure 3.6 An example (glucose) of what can happen if the application solvent (here H_2O) has not evaporated before elution. The right spot was applied first, then the plate was kept in a vacuum chamber for 2 minutes. The left spot was then applied and the plate was immediately eluted. The distortion of the left spot is caused by the remaining application solvent. When the elution solvent arrives at the spotted sample, the local solvent composition around it will vary and so will the mobility of the sample molecules, causing a "stretched-out" spot appearance.

treatment (Figure 3.6), or doing a complete "mini-workup" including quenching and liquid–liquid washing.

3.2.3.3 TLC Plate Elution

Maybe the most important thing when performing a TLC analysis is to choose a suitable elution solvent, and therefore a discussion of this subject will follow. However, first, some important technical things about the elution equipment: the elution starts when the TLC plate with the sample spots is put into the elution chamber with its bottom solvent layer. The chamber itself can be of a very simple design; it does not have to be a dedicated type that vendors offer (Figure 3.7a). A simple jar of suitable size with a lid works just fine (Figure 3.7b).

(a) (b)

Figure 3.7 (a) A commercially available elution chamber and (b) budget-type TLC elution chamber.

3.2.3.4 Choice of Elution Solvent

How to figure out a suitable elution solvent for your sample? The answer is: trial and error. However, there are so many solvent combination possibilities (5–10 pure ordinary lab solvents, and then thousands of binary, ternary, and even quaternary mixtures of these), so it helps if you have a rudimentary understanding of how solvent properties affect sample behavior on a TLC plate. A thorough presentation of TLC theory is beyond the scope of this text (consult Stahl's books [1, 2] and their references), but here is how you can think when trying out silica plate elution solvents: silica gel (the stationary phase) is a polar material, with polar groups extending from the particle surface. This means that the sample retarding forces are of polar nature, such as charge–charge interactions and/or hydrogen bonds. Highly polar samples experience stronger retardation and travel slower up the plate than less polar ones, regardless of the elution solvent. In fact, it takes a highly polar solvent to make a highly polar sample migrate at all. What the polar solvent does then is to partially block the polar sites on the silica gel surface, "protecting" the sample from the retarding forces. So, in a given binary solvent mixture, increasing the percentage of the polar component increases the migration speed of most compounds. This is a simplified picture of what is going on, but it is good enough to be useful in the selection of a good elution solvent. In Figure 3.8, you can see how increased solvent polarity increases the migration speed (R_f value) for both sample spots. You can also see that the separation of the two compounds is best at 20% and 40% ethyl acetate, where the R_f values are intermediate. This illustrates the

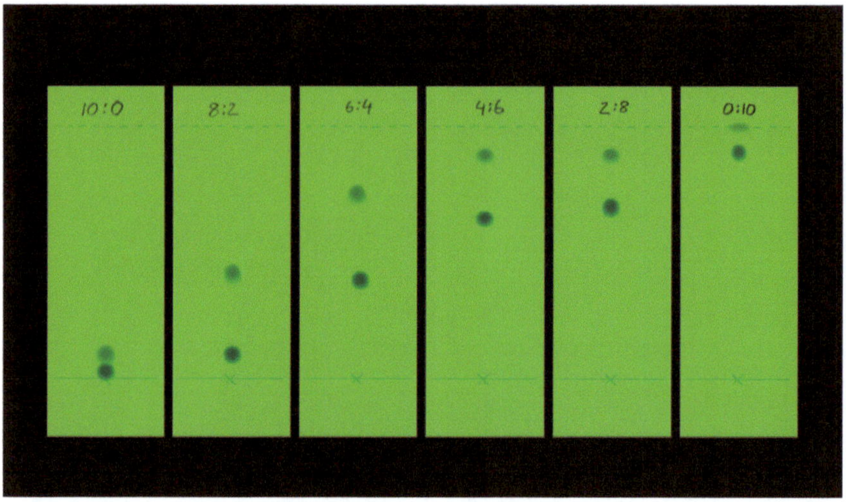

Figure 3.8 A 1:1 mixture of two aromatic compounds (4-nitro-*N*-trifluoroacetylaniline and 4-amino-*N*-trifluoroacetylaniline) eluted with different toluene–ethyl acetate mixtures (from 100% toluene to 100% ethyl acetate in five steps, UV visualization). Ethyl acetate is more polar than toluene, so the solvent polarity increases from left to right.

general rule that the spots should be near the middle region of the plate for the best separation.

With the above considerations in mind, you can start experimenting with solvents and solvent mixtures, moving from one end of the polarity spectrum to the other (usually from low to high polarity). Below is a list of common TLC elution solvents in an approximate order of increasing polarity. Note that the last four solvents are primarily used as additives to less polar solvents, not as single elution solvents (pure water, for example, makes the silica layer very unstable):

Pentane → hexane → heptane → toluene → chloroform → dichloromethane → diethyl ether → ethyl acetate → acetonitrile → acetone → isopropanol → ethanol → methanol → acetic acid → water.

3.2.3.5 Special Considerations Regarding the Choice of Elution Solvent for Highly Polar, Acidic or Basic Samples

Most organic chemists who use TLC are working at the lower end of the elution solvent polarity spectrum. This is because they normally study compounds of low or intermediate polarity and use silica plates. A typical example is the aromatic compound mixture in Figure 3.8. Now, what about really polar molecules, like carbohydrates, amino acids, or nucleic acids, can they also be analyzed/separated by TLC? The answer is yes, even potentially charged molecules like amines or carboxylic acids can be made to behave well on silica TLC plates. However, elution solvent mixtures of high polarity must be used, and it helps if they also contain a "buffering" component. The rationale for the presence of a buffer is that acidic or basic compounds need to be kept locked in one of their states. The two states differ greatly in TLC mobility and an ongoing equilibrium between the two states during elution can give very distorted and stretched-out spots and negligible spot–spot separation. A buffer favoring one of the states can improve the situation. Many useful solvent mixtures/buffers have been suggested; two examples (there are many others) are discussed below:

- For carbohydrates and acidic compounds:
  Ethyl acetate–methanol–acetic acid–water, *X*:3:3:2, where *X* is varied from 80 down to 4.
- For carbohydrates and basic compounds:
  Isopropanol–water–concentrated aqueous ammonia, *X*:4:1, where *X* is varied from 20 down to 8.

In Figure 3.5, you can see what can be achieved with silica plates and the polar system ethyl acetate–methanol–acetic acid–water (10:3:3:2). Figure 3.9 illustrates the importance of a buffering component in the elution solvent when a potentially charged compound (here a carboxylic acid) is analyzed by TLC.

3.2.3.6 Visualization by Exposure to UV Light

In Section 3.2.2, the first attempt to visualize colorless compounds was with UV light. If a compound absorbs UV light, it can be visualized by looking at the eluted and dried plate under a UV lamp (Figure 3.4i). The porous layer in most TLC plates contains a fluorescent additive that makes the plate shine greenish under a 254 nm ("UV-C") lamp. Compounds that absorb at 254 nm will then appear as dark spots (Figures 3.8–3.10). Many aromatic compounds absorb at this wavelength, albeit with different "efficiency" (i.e., their molar absorption coefficients at 254 nm differ). Therefore, you have to be careful when estimating concentrations from UV-spot intensities: a mixture with one strong and one faint spot may actually contain more of the faint-spot compound than you think. Figure 3.10 illustrates such a situation.

3.2.3.7 Visualization by Staining

The second attempt to visualize colorless compounds in Section 3.2.2 was by staining (Figure 3.4j,k). There are many known reagents that "stain" colorless organic compounds, and some of them have a functional group specificity. For example, compounds with amino or amido groups are stained purple-reddish with the ninhydrin reagent. Other reagents have a broader specificity: most carbohydrates, as well as many other organic compounds, are stained black by heating with sulfuric acid ("charring"), for example.

The TLC plate staining procedure typically involves gentle blowing/heating of an eluted plate (to remove remaining elution solvent), quick dipping in a solution of a staining reagent, and heating/blowing to induce the staining reaction. The staining solution can also be applied to the plate by spraying instead of dipping. However, this requires a dedicated "spray hood" or "spray box," since the staining reagents are often toxic and/or corrosive.

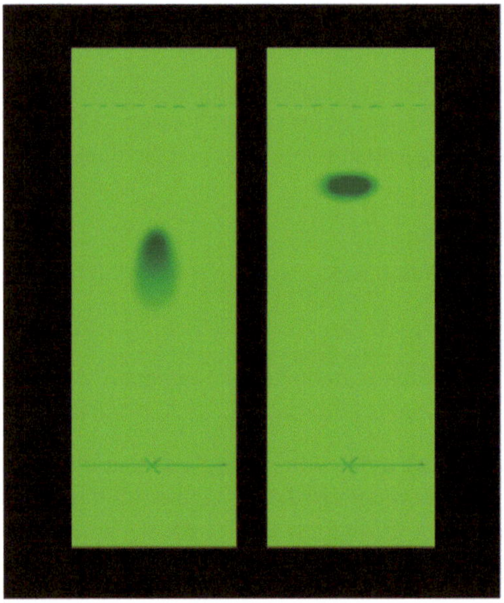

Figure 3.9 4-Methoxybenzoic acid in 1:1 toluene–ethyl acetate without (left plate) and with (right plate) added 5% acetic acid. UV visualization.

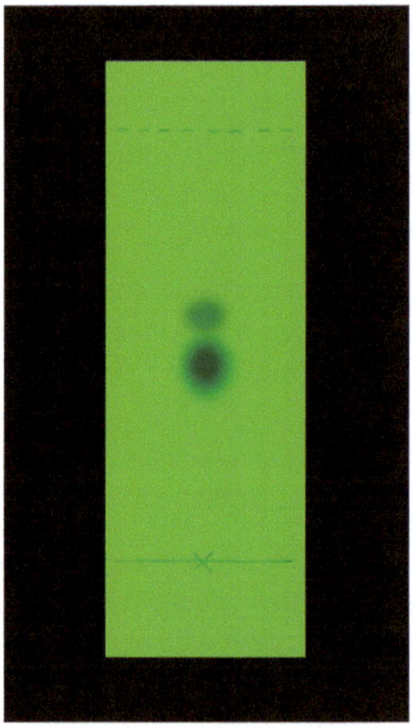

Figure 3.10 TLC chromatogram of a 1:1 (0.05 μmol each) mixture of 4-aminophenol (faster spot) and 4-acetamidophenol (slower spot). Elution solvent: ethyl acetate. UV visualization.

Several hundred different staining reagents and procedures are known and described, and it can be difficult for a beginner to know which ones are best suited for his/her compounds. The recommendation here is to start with some of the more general reagents that stain broadly such as sulfuric acid (with or without additives), phosphomolybdic acid, ninhydrin, or iodine (the latter is applied as vapor in a closed chamber, not sprayed or dipped). The details of how to prepare and use these four reagents are given below. For more details on staining reagents, see, for example, Stahl's handbooks [1, 2], which have a list of several hundred staining reagents. It cannot be enough stressed: good knowledge of staining procedures is of great help in TLC analysis because it facilitates fast visualization/identification of compounds in a mixture, and can also provide useful information about functional groups.

3.2.3.8 Four Useful General Stains (Recipes and Comments)

Sulfuric acid: prepare a 5% (v/v) solution of sulfuric acid in water or ethanol. Dip or spray, then heat with a heat gun or hotplate. Most carbohydrates, and many other organic compounds as well, stain black or brown. The stain mostly consists of "carbonized" organic matter. There are many varieties of this staining procedure where additives are used to enhance/colorize the spots or broaden the specificity. One is *anisaldehyde–sulfuric acid*: anisaldehyde (0.5 mL) is mixed with acetic acid (10 mL), followed by methanol or ethanol (85 mL) and concentrated sulfuric acid (5 mL).

Phosphomolybdic acid: dissolve phosphomolybdic acid (1–10 g) in 99% ethanol (100 mL). Dip or spray, then heat with a heat gun or hotplate. The plates will appear greenish, with darker green or blue–green spots. Many different types of compounds stain with this reagent.

Ninhydrin: prepare a 0.3% (w/v) solution of ninhydrin in ethanol or 1-butanol containing 3% (v/v) acetic acid. Dip or spray, then heat with a heat gun or hotplate. Amines stain quickly with a bright purple color, amides stain slower (see, e.g., Figure 3.41). Many other compounds also stain slowly with ninhydrin, usually with a more brownish color.

Iodine vapor: make a chamber as follows: a few crystals of iodine and a filter paper are added to a 100–200 mL wide mouth jar (with cap). After a while, the chamber will become saturated with iodine vapor. Insert your TLC plate and allow it to remain within the chamber until it develops a light brown color over the entire plate. Commonly, if your compound has an affinity for iodine, it will appear as a dark brown spot against a light brown background. Carefully remove the TLC plate and mark the spots with a pencil and/or take a photograph. The iodine will not remain on the TLC plate for long so documentation has to be done rapidly.

3.2.3.9 TLC–Mass Spectrometry

TLC is an excellent tool for separating organic compounds. However, it gives you only hints (R_f values, UV absorbance, and functional group staining) about the identity and structure of the compounds in the spots. To get more specific information, you need to combine TLC with a modern analytical technique such as mass spectrometry or NMR spectroscopy. A typical spot on a 0.25 mm analytical TLC plate contains 1–5 μg of the material, which is more than enough for a good mass spectrum of any type. The most obvious way then is to scrape off the solid-phase layer under the spot (visualized by UV light or by staining of a parallel plate), extract it with a solvent, filter off the solid, and get a mass spectrum of the filtrate. Figure 3.11 shows the equipment needed for this manual "extraction" procedure.

Much development resources have been spent to combine TLC and mass spectrometry in one instrument. So far, the most common technical solution is to let the instrument do the spot extraction, filtration, and subsequent analysis. However, the elegance of the electrospray LC–MS solution, where the column eluate is directly "sprayed" into the mass spectrometer, has yet to be matched with TLC. There are ongoing attempts to use new ways to transfer material from TLC plates into a mass spectrometer.

A word of caution concerning the extraction of compounds from silica TLC plates (regardless of whether it is done manually or automatically by an MS instrument): the procedures work best with compounds that are of low or intermediate polarity. These compounds extract well from straight-phase silica with solvents like acetonitrile or acetone. In contrast, polar compounds need highly polar extraction solvents (e.g., alcohols) to "release" from the silica, and problems can then arise with co-extraction of unwanted material from the plate (ions, additives). Such contaminants

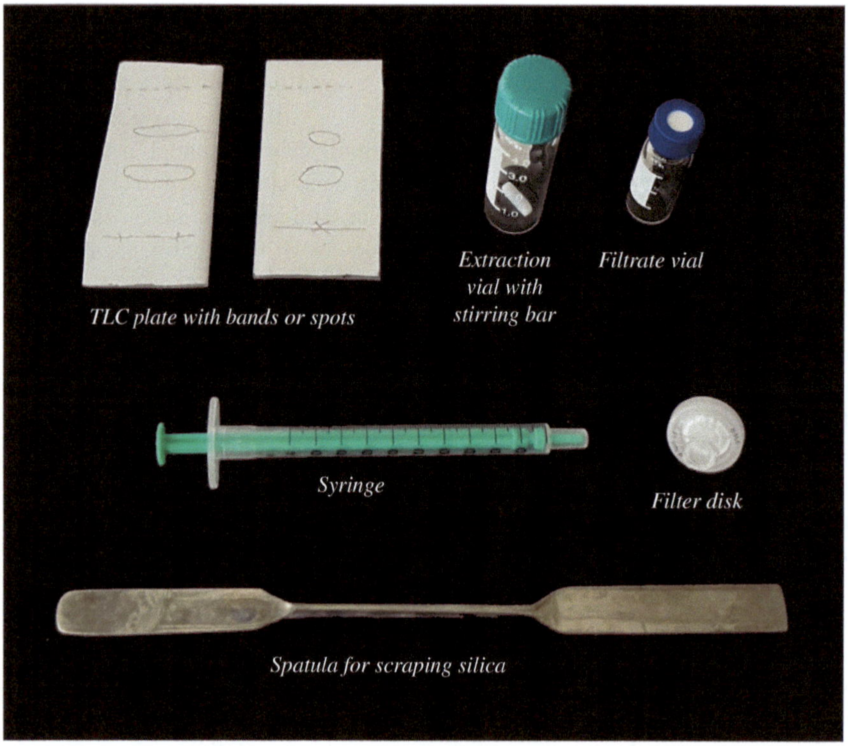

Figure 3.11 Equipment for manual TLC extraction. The amount of material in a spot (1–5 μg) is usually enough for a mass spectrum, but to play it safe one could apply a band instead of a spot on the plate, this will give more material.

can "saturate" a mass spectrum with irrelevant signals, e.g., from sodium or ammonium ions. Therefore, a "TLC–MS spectrum" from a polar compound may not show you much of the expected ions, even though the compound is actually there in the spot. The compound might not have extracted well from the silica, or it did so but together with contaminants that obscured the mass spectrum. Some TLC plate vendors have tried to address these problems by providing special "MS-grade" TLC plates, with increased sensitivity and less potential contaminants. Time will tell how much of an improvement these new plates can bring (in relation to their higher cost).

3.2.3.10 TLC–NMR

A standard 400 MHz instrument with an ordinary probe requires 1–10 mg for a satisfactory H-NMR spectrum, so to extract such amounts from a 0.25 mm TLC plate it has to be a large one (preferably the standard 20 × 20 cm size), and the sample should be applied as a band. It may or may not be practical to scale up a TLC analysis this much, but with an organic reaction in the 200–2000 mg scale it is usually possible. The TLC–NMR combination works well in such cases. It is performed in many labs as a fast and cheap alternative to running a small chromatography column or semi-preparative HPLC to get pure samples for NMR analysis.

3.3 Liquid Chromatography–Mass Spectrometry

For many decades, TLC was the dominating method for reaction monitoring in organic chemistry. However, when reliable LC–MS instruments became available, some chemists started to use them for reaction monitoring. The advantages were obvious: the components of a reaction mixture could not only be separated (as with TLC) but also be identified by their molecular mass. The analysis time and the amount of sample needed were similar to those of a TLC analysis.

Today, LC–MS is an established alternative to TLC for reaction monitoring. However, it is *much* more expensive to monitor a reaction with LC–MS than with TLC, and LC–MS is also more sensitive to disturbances: if an LC–MS instrument is malfunctioning or down for maintenance, it can affect many users. TLC is simpler and more robust, in comparison.

3.3.1 LC–MS Introduction

A short description of a typical LC–MS instrument follows. For more details, see reviews [5, 6].

A standard LC–MS instrument (Figure 3.12) consists of a pump/gradient mixer, an autosampler (or a manual injection port), an HPLC column (most often of the silica gel reversed-phase type), a UV detector, a mass spectrometer/detector, and a computer.

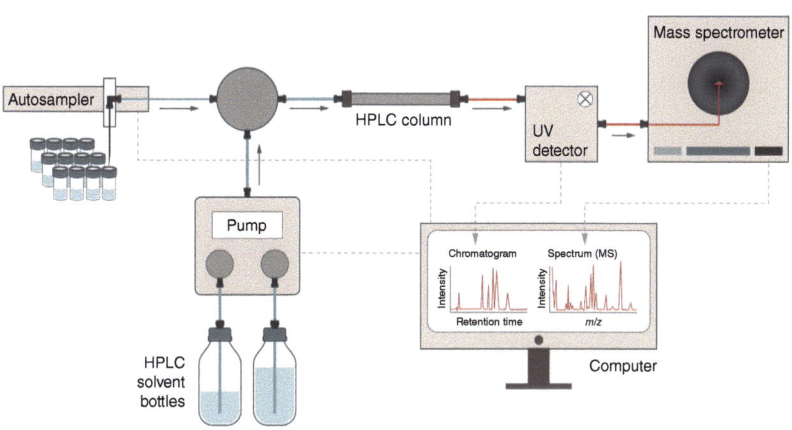

Figure 3.12 Important components of an LC–MS instrument. Courtesy of Margaret Holtz.

The workflow for a single sample analysis is as follows:

- Make a 0.5–2 mg/mL sample solution (preferably filtered, see Chapter 5, Figure 5.4) in a standard HPLC vial and place it in the rack of the autosampler (Figure 3.12).
- Enter the sample coordinates and other instrument settings in the appropriate computer windows and execute a "run" command. The autoinjector then delivers the sample solution to the HPLC column. The components separate and exit the column at different times and then enter the UV and mass detectors.
- The UV detector senses the UV absorption(s) and records absorption versus time chromatogram(s).
- The mass detector, usually of the ESI ("Electrospray Ionization") type, senses the ions passing through, and records the "total ion current" (TIC) versus time chromatograms (positive ions and negative ions). These chromatograms are called TICs.
- The mass detector *also* produces mass spectra (ion intensity versus m/z diagrams) for all data points in the +TIC and −TIC chromatograms (amounting to quite a few spectra).
- Data from both detectors are collected and organized by the computer. Figure 3.14 illustrates a common data display format: a UV chromatogram is shown together with its (almost) simultaneously recorded +TIC chromatogram (shown here on top of each other in the left half of the screen). By selecting a data point in the TIC chromatogram on the screen, you can "extract" the mass spectrum recorded for that point (shown here in the left half of the screen). You can also extract mass spectra for other points in the TIC chromatogram, like in Figure 3.15.
- The mass spectral information (the molecular ion and fragment ions) provided for each peak is often enough to identify its content, a remarkable feat considering the short time and tiny sample amount.
- In addition, instruments with multiple-wavelength UV detectors ("DADs": diode array detectors) produce UV spectra for each data point in the UV chromatogram, providing further component information.

3.3.2 LC–MS Monitoring Versus TLC Monitoring, Real-world Example

LC–MS monitoring of the reaction in Figure 3.13 is shown in Figures 3.14 and 3.15 and compared to monitoring by TLC (Figure 3.16).

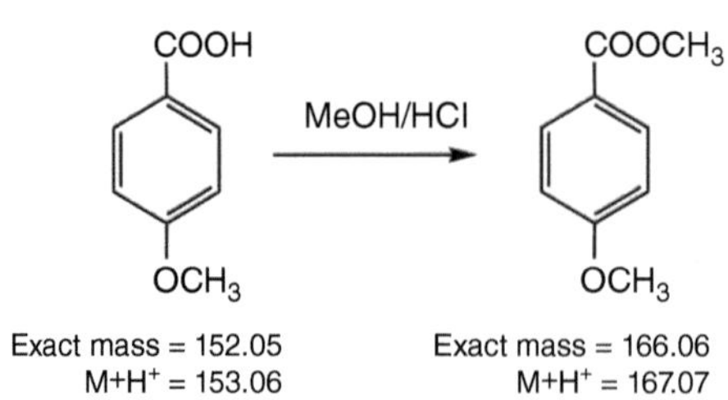

Figure 3.13 Example reaction: hydrochloric-acid-catalyzed esterification of 4-methoxybenzoic acid in methanol solution.

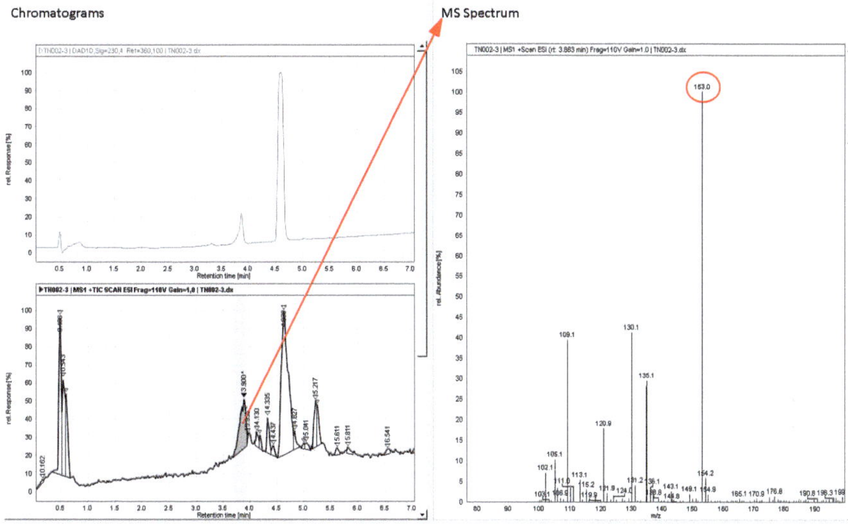

Figure 3.14 Left: two LC–MS chromatograms (UV and +TIC) of the reaction mixture. The arrow in the +TIC chromatogram indicates where the mass spectrum was extracted. Right: the extracted MS spectrum (the M + H ion is encircled).

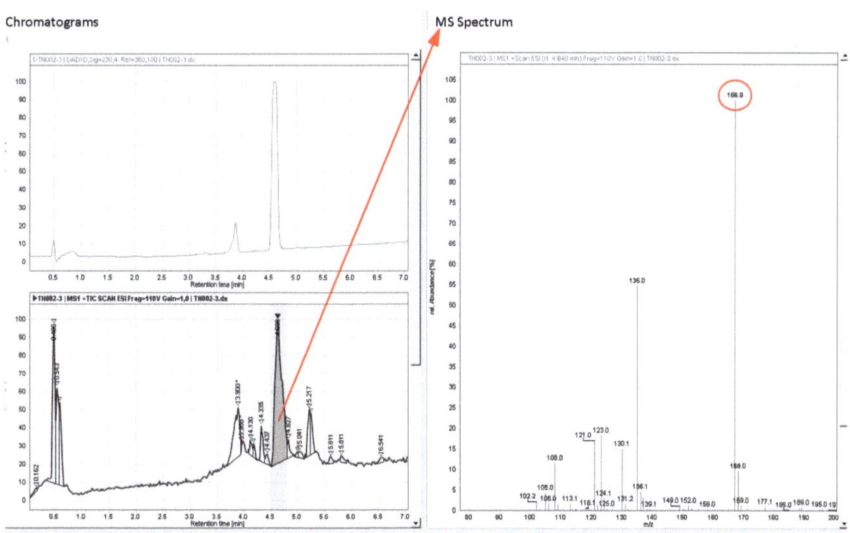

Figure 3.15 Left: the same two LC–MS chromatograms (UV and +TIC) of the reaction mixture. The arrow in the +TIC chromatogram indicates where the mass spectrum was extracted. Right: the extracted MS spectrum (the M + H ion is encircled).

The reaction was run at room temperature with 20 mg of p-methoxybenzoic acid (PMB) in 10 mL of methanol to which was added 200 μL of acetyl chloride (this is a convenient way to generate HCl/methanol). Being reasonably dilute (2 mg/mL) and relatively "pure" from salts and other additives, this mixture was analyzed directly,

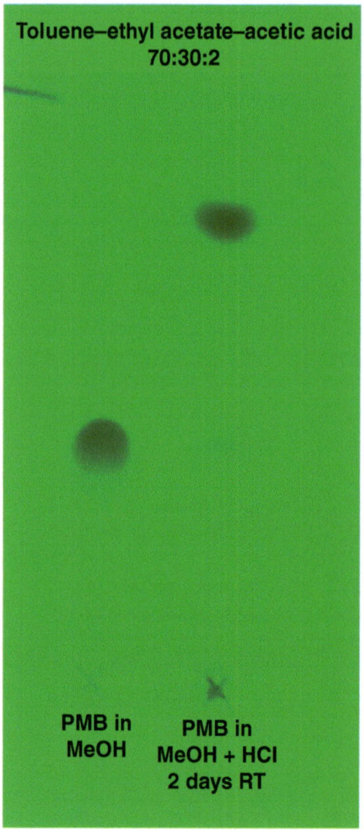

Figure 3.16 TLC chromatogram of PMB in methanol (left lane) and the reaction mixture (right lane). Visualization was by 254 nm UV light.

without dilution or preprocessing. Aliquots of the reaction mixture were withdrawn after two days of reaction:

LC–MS: first, 10 μL was injected into an LC–MS equipped with a short C-18 column and a 10–90% water–acetonitrile gradient containing 0.1% formic acid. UV detection was at 230 nm.

TLC: then 2 μL was spotted onto a silica gel TLC plate next to a PMB-in-MeOH spot, and the plate was eluted with a toluene–ethyl acetate–acetic acid (70:30:2) mixture. UV visualization was at 254 nm.

The results are shown in Figures 3.14–3.16.

In this example, both LC–MS and TLC showed that the reaction had *not* reached completion, since there was some starting material left in the mixture. Note that the product ester, which is less polar than the starting carboxylic acid, was the "slow" component on the LC and the "fast" component on the TLC chromatograms. This is because LC used reversed-phase silica (lower polarity equals slower migration) and TLC used straight-phase silica (lower polarity equals faster migration).

The mass spectra corresponding to the first and second UV peaks showed the expected $[M + H]^+$ molecular ions ($m/z = 153$ and 167), confirming that these peaks were from starting material and product, respectively. In addition, both mass spectra showed an $m/z = 135$ ion. This is most likely a fragment ion from "alpha-cleavage" of the O═C—O single bond present in both compounds.

The relative peak intensities shown in the +TIC chromatogram differ a little from what you see in the UV chromatogram. If the UV absorbances are comparable, the +TIC difference could be because different compound classes ionize better than others in a mass spectrometer. If the LC–MS mass detector is of the ESI type, polar compounds (alcohols, ketones, aldehydes, amines, amides, esters, carboxylic acids, etc.) ionize well, especially as positive ions, e.g., $[M + H]^+$, $[M + Na]^+$. Carboxylic acids ionize well as negative ions, e.g. $[M - H]^-$. To improve ionization, buffers are often added to the LC eluant (TFA, formic acid, or ammonium salts). Buffers promote the formation of charged adducts like $[M + H]^+$, $[M + Na]^+$, and $[M + NH_4]^+$. Buffers can also greatly improve HPLC separations.

Many nonpolar compounds (hydrocarbons, lipids, and steroids) do not ionize well in an ESI mass detector. Such compounds might even pass an ordinary LC–MS undetected, especially if the UV absorbance is low. LC–MS analysis of nonpolar compounds often works better with an APCI mass detector.

Finally, note that the TLC chromatogram (Figure 3.16) shows *three* UV-absorbing spots (in the order of decreasing R_f value): the product material, the starting material, and a spot that had not migrated upward at all ($R_f = 0$, "starting spot"). Starting spots usually indicate that some degradation into highly polar compound(s) had occurred during the reaction. This is important information, since it could explain an unexpectedly low product yield. The polar degradation compound(s) went undetected in the LC–MS chromatograms since they eluted together with the "front peak"! This is a common phenomenon and should be remembered.

3.3.3 LC–MS Tips and Tricks

LC–MS instruments come with different sensitivities and configurations; therefore, the sample volume/concentration used in the above example may not be optimal on other instruments. Unexpectedly low peak intensities in the chromatograms and "noisy" mass spectra indicate that the sample volume/concentration is too low and should be increased. Conversely, ugly-shaped UV chromatograms and positive ion mass spectra with strong 2M + H ("dimer") peaks indicate that the sample volume/concentration is too high and should be decreased.

As regards the mass accuracy of a routine instrument, you should trust only the numbers before the decimal point. This means that, as in the mass spectrum in Figure 3.15, the m/z value 166.9 should be interpreted as 167 (with no decimal point). There are ways to fine-tune instruments for higher accuracy, but it is seldom needed for reaction monitoring.

3.3.4 LC–MS Versus TLC Monitoring

Compared to LC–MS monitoring, TLC monitoring gives an often faster and better "overview" of mixture components since all possible R_f-s are covered, from 0 to 1 (provided that all spots can be visualized). During LC–MS monitoring, you often "miss" compounds with extreme R_f-s, but on the other hand you have the huge advantage of getting detailed structural information directly about all compounds that do show up as peaks. TLC monitoring does not provide such information directly, but it can be obtained by some extra efforts in many cases (TLC–mass spectrometry or TLC–NMR), see Sections 3.2.3.9 and 3.2.3.10 for details.

3.4 Monitoring of Solid-Phase Reactions

A solid-phase reaction cannot, like a solution-phase reaction, be monitored directly by TLC or LC–MS since the reactant and product are both anchored to a solid phase and not in solution. However, indirect methods can be used, here are some examples:

1. During a manual solid-phase peptide synthesis, the amino acid coupling efficiency can be estimated by removing a small aliquot of the solid phase after the coupling and subjecting it to the "Kaiser" ninhydrin test [7], which produces an intense blue color if there are any free amino groups left on the solid phase. A negative test indicates a complete coupling reaction. This is a qualitative test only.
2. Another, more elaborate way to monitor a manual peptide synthesis is to remove a small aliquot of the solid phase after a reaction, cleave off the anchored peptide into solution, and analyze the solution by LC–MS or MALDI–MS. This also is a qualitative test only.
3. A coupling yield in solid-phase peptide syntheses that use the Fmoc temporary N-protection group can be monitored by measuring the UV absorbance of the combined filtrate and washings obtained from the subsequent Fmoc-deprotection step. This step removes the Fmoc protecting group from the solid-phase-anchored peptide, thereby releasing into solution a UV-absorbing dibenzofulvene compound. The amount of this compound can be calculated from the UV absorbance of the Fmoc-deprotection filtrate. Assuming that the Fmoc-deprotection step was quantitative, this amount should reflect the amount of coupled amino acid in the preceding step, and thus the coupling yield.
4. In a similar way, solid-phase nucleotide syntheses that use trityl-type temporary protecting groups can be monitored by measuring the absorbance of deprotection filtrates. Here, the red or orange color from released trityl compounds can even be seen with the naked eye. Figure 3.17 shows the filtrate and washings collected in measuring flasks during a manual nucleotide synthesis, one flask for each deprotection reaction. Monomethoxytrityl temporary protecting groups were used here.

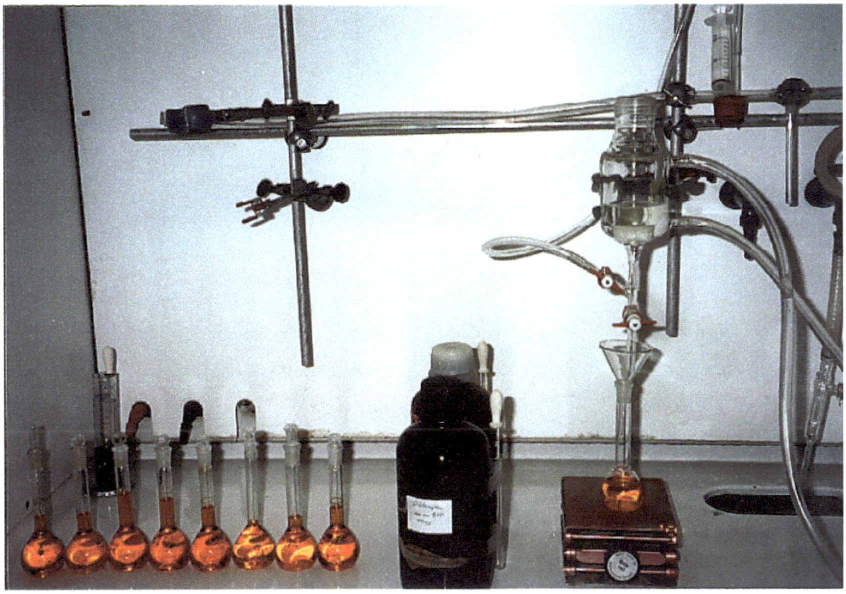

Figure 3.17 Filtrate and washings collected in measuring flasks during a manual nucleotide synthesis, one flask for each deprotection reaction. The reaction vessel is seen in the upper right.

References

1 Stahl, E., *Dünnschicht-Chromatographie: Ein Laboratoriumshandbuch*. 1967, Berlin: Springer-Verlag. 979 pages.

2 Stahl, E., *Thin-Layer Chromatography: A Laboratory Handbook*, 2nd Ed. 2012, Berlin: Springer-Verlag. 1041 pages.

3 Fried, B. and J. Sherma, *Practical Thin-Layer Chromatography: A Multidisciplinary Approach*. 1996, New York: CRC Press.

4 Leonard, J., B. Lygo, and G. Procter, *Advanced Practical Organic Chemistry*, 3rd Ed. 2013, Boca Raton, FL: CRC Press. 356 pages.

5 McMaster, M.C., *LC/MS: A Practical User's Guide*. 2005, Hoboken, NY: Wiley-Interscience. 157 pages.

6 Kind, T. and O. Fiehn, Advances in structure elucidation of small molecules using mass spectrometry. Bioanal Rev, 2010. 2: pp. 23–60.

7 Wellings, D.A. and E. Atherton. Standard Fmoc protocols. In G.B. Fields, editor, *Solid-Phase Peptide Synthesis*, volume Methods in Enzymology, Chapter 4, pages 44–67. Academic Press, 1997.

4

Reaction Workup

4.1 General Considerations

When a reaction is "finished" (nothing more seems to happen in the flask, and a quench reagent has been added if necessary), you have to isolate the reaction product from remaining starting material, by-products, additives, and solvent(s). This is called "working up" the reaction mixture (see the workflow scheme in Figure 1.1 in Chapter 1). The first operation is often a liquid–liquid wash ("aqueous workup," Section 4.2), a filtration (Section 4.3), or both. After an eventual evaporation (Section 4.4), a final purification is performed. This could be a distillation (Section 4.5) or a crystallization (Section 4.7), but most often nowadays it is some kind of column chromatography (Section 4.8) followed by a final evaporation (Section 4.4) or freeze-drying (Section 4.6) of the pure product(s).

4.2 Liquid–Liquid Washing

4.2.1 Introduction

This very common procedure is used to remove one or more polar components from a reaction solution. When a diethyl ether, toluene, ethyl acetate, or dichloromethane solution is added to water (or dilute acids/bases), there will be two layers ("phases") formed no matter how you stir or shake. This is because water and the organic solvents have limited solubility in each other (they are "immiscible"). The phenomenon is used to selectively transfer ("wash out" or "extract") salts or other polar compounds from the reaction solution ("organic phase") to the water phase. Polar compounds (like salts) simply prefer water over organic solvents, and they move over to the water phase if you shake a two-phase system long enough. Conversely, less polar compounds prefer to stay in the organic phase. If you, after shaking, separate the two phases, you have the salts and other polar compounds in the water phase and the less polar compounds (the reaction products) in the organic phase (Figure 4.1).

Figure 4.1 Two phases physically separated by liquid–liquid washing.

The three most common liquid–liquid wash types are:

1. *Acidic aqueous wash*: the water phase is a dilute acid and the organic phase is any immiscible organic solvent. This wash removes basic compounds from the organic phase by converting them into polar, water-soluble salts.
2. *Basic aqueous wash*: the water phase is a dilute base (sodium bicarbonate, carbonate, or hydroxide) and the organic phase is any immiscible organic solvent. This wash removes acidic compounds from the organic phase by converting them into polar, water-soluble salts.
3. *Aqueous wash or brine wash*: usually done as a last wash to free an organic phase of the remains from previous aqueous washings. "Brine" is saturated aqueous sodium chloride.

The equipment needed for a typical liquid–liquid wash procedure is shown in Figure 4.2. A "step-by-step" description follows.

Figure 4.2 Equipment that you need for a liquid–liquid "wash": a stand with a ring or a clamp, a separatory funnel with a stopcock, an ordinary funnel, and some Erlenmeyer flasks.

4.2.2 Liquid–Liquid Washing, Step by Step (Figure 4.3a–g)

Figure 4.3a Pour the organic solvent solution and the aqueous solution into the separatory funnel (make sure that the bottom tap is closed when you do this). Use an ordinary funnel to avoid spills.

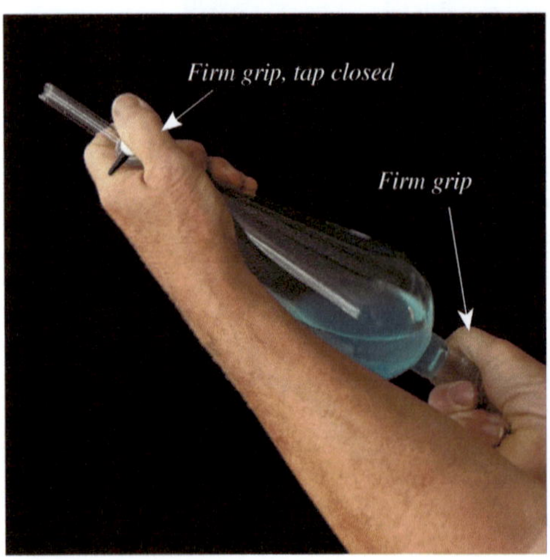

Figure 4.3b Shake the funnel (first time very briefly and gently!) while holding the stopcock firmly closed with one hand and the closed bottom tap with the other hand.

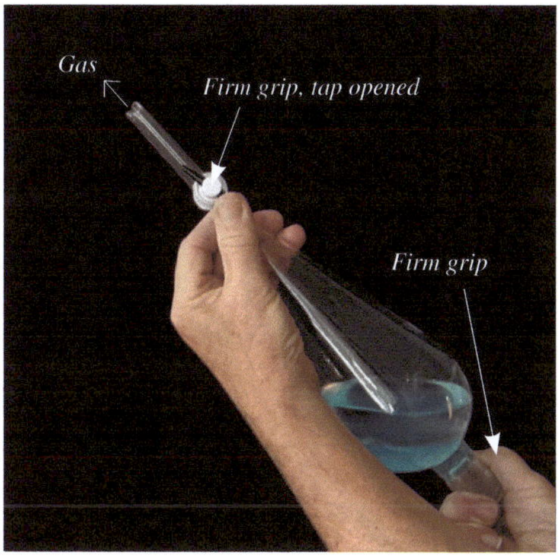

Figure 4.3c Immediately after the "shake," turn the funnel upside down and vent any eventual overpressure (see Section 4.2.3) by opening the bottom tap. The shaking/venting is repeated (with increasingly forceful shakings) several times.

Figure 4.3d,e Put the separatory funnel upright in the stand and remove the stopcock. Place an Erlenmeyer flask directly underneath and wait until the phases are well separated. Then open the bottom tap slowly and empty the lower phase *only* into the Erlenmeyer flask. Close the tap immediately before the phase interface enters it.

Figure 4.3f Pour out the upper phase into another Erlenmeyer flask.

Figure 4.3g You now have the two phases in separate flasks! In his example, the organic (blue) phase is dichloromethane, which is denser than water and therefore will be the lower phase in the separatory funnel.

4.2.3 Washing Tips/Tricks

1. *Safety considerations*: it is rather difficult to avoid solvent spills when you perform a liquid–liquid wash. The greatest risk is that an overpressure in the separatory funnel causes a "splash-out" through the stopcock, read more about this below. Safety goggles and a lab coat are *mandatory*, and you should do everything in a fume hood! You could also consider using gloves. If you do not use gloves you should always rinse your hands directly afterward, even if you did not notice any spills (small acid and base spills can go unnoticed but are felt after a while).

2. *Overpressure venting and pH check*: a common mishap during an acidic aqueous wash with an aqueous bicarbonate/carbonate solution is the overpressure "splash-out" that can occur if shaking and venting is not done properly. Carbonates quickly release lots of CO_2 in contact with acids, and the pressure build-up inside the closed separatory funnel is fast. It is good practice to first shake *very* shortly and gently, and immediately vent the overpressure, then shake again, vent, and so on. When no more gas is released during venting, make sure (with indicator paper) that the pH of the aqueous phase is well above 7, indicating that all eventual acids in the organic phase have been neutralized and washed out (which is what you want). Finish with a final vigorous shake.
3. *Small-scale Washings*: Separatory funnels come in many different sizes (several L down to 10 mL). In the lower volume range (<50 mL), liquid losses during washings become significant, therefore a simpler setup (Figure 4.4) is an alternative. The principle is the same, but the lower phase is not drained with a bottom tap, instead it is withdrawn from the bottom of a pear-shaped flask with a Pasteur pipette.
4. *Working* "quantitatively": especially when you work on a small scale, there can be considerable losses of both phases (drops adhering to the walls of the separatory funnel, etc.). It is good practice to perform an extra "wash" of the separatory funnel (or pear-shaped flask) with a small amount of fresh lower-phase liquid directly after the first layer separation. You then combine the separated "wash" layer with the first separated layer.
5. *Which layer is the* "organic phase"?: Diethyl ether, toluene, and ethyl acetate solutions are less dense than water and will therefore end up as the upper layer in a liquid–liquid extraction. Dichloromethane and chloroform solutions are denser than water, and will therefore end up as the lower layer. However, when solvent mixtures or salt-rich water solutions are involved, the layer order can actually be quite difficult to predict. A simple test will then clarify: addition of an extra amount of one solvent will increase the volume of its corresponding layer!
6. *Emulsions*: An irritating and common phenomenon when doing washings of any kind is that the phases do not separate well after shaking. A milky emulsion is all you see inside the separatory funnel, no matter how much you wait for the g-forces to separate the phases. You can then try several "cures," not necessarily in the order presented here: (i) if you are not in a hurry, you can simply leave the separatory funnel in its stand overnight, and hope for a better appearance the next morning. (ii) If you cannot wait that long and your emulsion has a rather small (10–100 mL) volume, you can transfer it to a (closed) centrifuge tube and spin with a lab centrifuge (do not forget the counterweight), thereby increasing the g-forces and thus speeding up the phase separation. (iii) You can try dilution of one phase, or both. If this does not work and the organic phase is the upper one, you can try dilution with a solvent of very low viscosity and density, like diethyl ether. (iv) Increasing the ionic strength of the aqueous phase can also sometimes help. This can be done by adding brine (i.e., a saturated solution

Figure 4.4 Alternative setup for small-scale (<50 mL total volume) washings.

of sodium chloride in water), or even some solid sodium chloride. (v) In cases when low pH is not an issue, you can try to add some acid. (vi) You can try filtering the entire emulsion. It sometimes helps, especially if the filter is of the phase-separating type.

7. *The last step, organic phase drying*: In most liquid–liquid washings, the organic phase contains the reaction product(s), and further manipulations with this solution will therefore follow. Removal of the remaining moisture is often the first step ("drying"). You do this by adding a solid drying agent (e.g., anhydrous magnesium or sodium sulfate), stirring for a short period, filtering, and collecting the filtrate (see the step-by-step procedure in Section 4.2.4).

4.2.4 Organic Phase Drying, Step by Step (Figure 4.5a–c)

Figure 4.5a Add a magnetic stirring bar to the flask with the organic phase, followed by anhydrous magnesium sulfate or anhydrous sodium sulfate. Swirl gently. If the solid clumps together, add more until it "snows" in the flask.

Figure 4.5b Cover the flask loosely with aluminum foil or an inverted beaker and stir for 5–10 minutes. The liquid should soon become clear after the stirring has stopped. If not, add more drying agent and stir again.

Figure 4.5c Filter off the solid, preferably using the above type of adapter that delivers the filtrate directly into a flask suitable for evaporation. Wash the flask and the solid with small portions of fresh solvent.

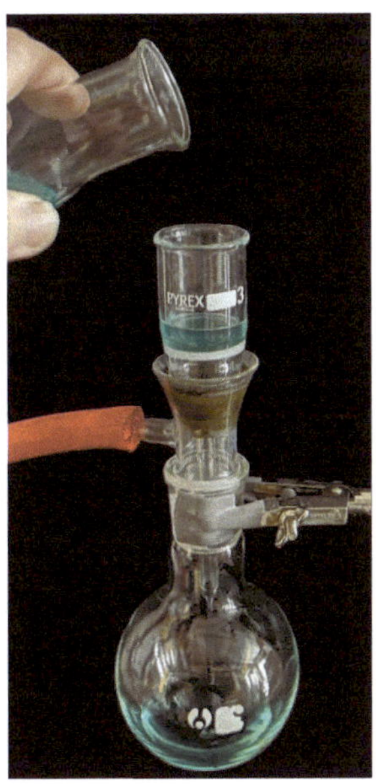

4.3 Filtration

4.3.1 Introduction

Filtration is one of the oldest separation techniques, it was practiced in kitchens and factories long before there were any chemistry labs. It separates a solid/liquid mixture into its components with the aid of a filter device. In today's milligram-to-gram-scale organic chemistry, the most common filter device is the fritted glass funnel (Figure 4.6, lower row). This funnel as well as its older relative (the Büchner funnel with a paper filter; Figure 4.6, upper row) are normally used in combination with a suction flask (Figure 4.7) where vacuum from a simple source can be applied to speed up the process. Typical setups for such "suction filtrations" are shown in Figures 4.8 and 4.9. Note that you should attach a safety flask (Figure 4.10) between the setup and the vacuum source to collect accidental spillovers.

Figure 4.6 Büchner funnel/paper filter and fritted glass funnels of different sizes.

Figure 4.7 Suction flasks of different sizes.

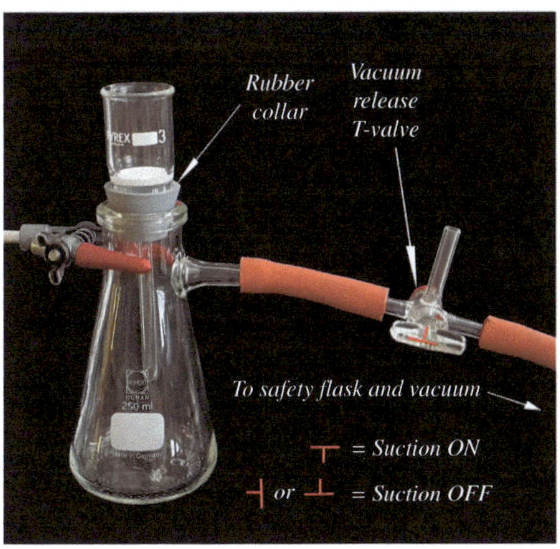

Figure 4.8 A typical suction filtration setup. You can omit the T-valve if the vacuum source has its own open/close switch.

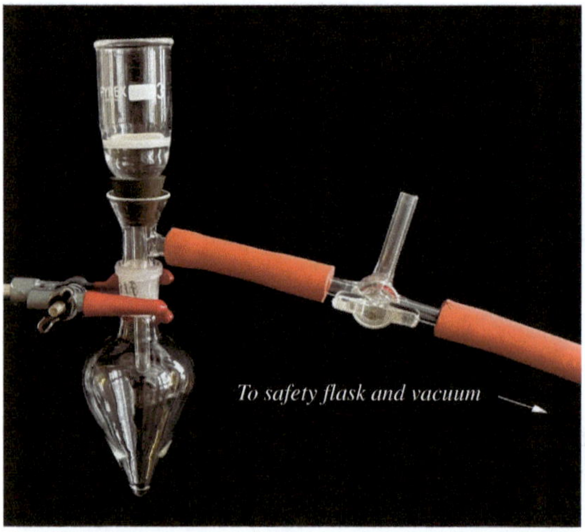

Figure 4.9 Alternative suction filtration setup where the filtrate is collected directly into an evaporation flask.

Figure 4.10 Example of a safety flask. It collects accidental spills/back-suctions from either direction.

4.3.2 Step-by-step Suction Filtration Where the Solid Is the Important Thing. This Example Is with Crystals from a Recrystallization, Figure 4.11a–f

Figure 4.11a With suction off, stir up the mixture (e.g., by swirling) and pour the slurry quickly onto the funnel. With a small funnel and a large amount of slurry you should pour portion-wise, with suction in between. For Büchner funnel use, see Section 4.3.3.1.

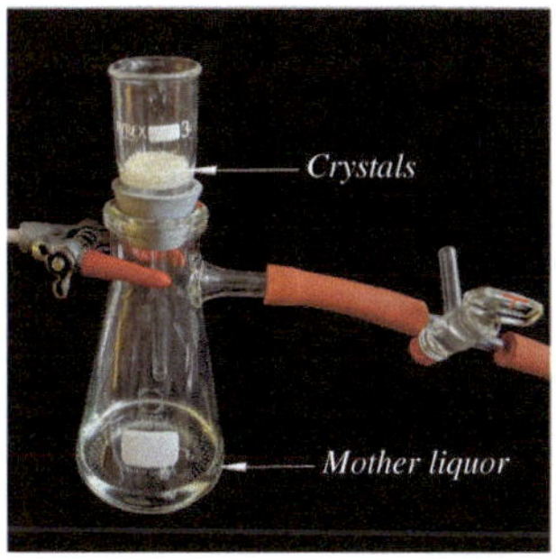

Figure 4.11b Apply suction until the crystals on the filter are almost dry, then stop the suction. The filtrate is here called the "mother liquor" (the crystals were "born" out of it).

Figure 4.11c Wash the crystals by adding a little fresh, cold solvent to the original flask, swirling, and pouring it onto the crystals on the filter. Let it penetrate thoroughly, then apply suction. Wait until no more liquid filters off. Repeat the process once more.

Figure 4.11d After the last wash, you can leave the suction on for some time to partially dry the crystals.

Figure 4.11e If you have tare-weighed your funnel before use, you can dry it with the crystals in, either in the open air or in a vacuum desiccator. Check the weight afterward and subtract the tare value.

Figure 4.11f Alternatively, transfer the crystals to a tare-weighed beaker or similar and leave them to dry. It is a good idea to cover the opening loosely with perforated aluminum foil during the drying process.

Depending on the purpose of the filtration, the procedure above can be modified, see Section 4.3.3.

4.3.3 Tips and Tricks

4.3.3.1 Büchner Funnel/Paper Filter

With a Büchner funnel, you should first wet the filter paper with a small amount of solvent and then apply suction to fix the paper snugly onto the funnel bottom. Do this immediately before filtering. The rest of the procedure is the same as with the fritted glass funnel.

4.3.3.2 Filter Aids

"Difficult" solid–liquid mixtures that are known to "clog" filters can be filtered using a filter aid (Celite Hyflo or similar) if only the filtrate is important. Filter aids are inert, fine-grained powders which themselves filter very well. First add a slurry of the filter aid to the filter funnel, then apply suction until a *wet* cake is formed. Carefully layer the "difficult" solid–liquid mixture on top of the filter-aid cake and apply suction again. If the filter-aid cake remains intact during the filtration, the result should be satisfactory and the process reasonably fast. Filter aids are often used to remove, for example, finely dispersed catalysts or silver halides (notoriously difficult to filter off!) from a reaction mixture.

4.3.3.3 Filtration of Small Amounts of Material Where the Solid Is the Important Thing

When the amount of solid (e.g., crystals) to filter off approaches single-digit milligrams, filtration becomes technically difficult and percentage losses increase. Below is a suitable procedure:

a. Use the smallest possible filter funnel, e.g., a 10-mm-diameter fritted glass funnel with porosity 4 (i.e., the far lower right one in Figure 4.6). Tare weigh the funnel dry and empty.
b. Cut the tip of a Pasteur glass pipette so that the opening widens enough to allow your solid/liquid slurry through.
c. Use this Pasteur pipette to transfer your slurry from the original flask to the small filter funnel and do a regular suction filtration + wash (rinse the original flask and the Pasteur pipette with the wash solvent).
d. Dry the funnel with the solid still in, and weigh. Subtracting the tare weight will give the yield.

Alternatively, you can isolate small amounts of solids by centrifugation + washing in tare-weighed Falcon or Eppendorf tubes.

4.3.3.4 Filtration of Small Amounts of Mixtures Where the Filtrate Is the Important Thing

You can clean up smaller (1–25 mL) amounts of solvent from finely dispersed solids with the aid of plastic filter cartridges and syringes ("sterile filters," Figure 4.12a,b). A very clean filtrate will be produced, however, the solid becomes trapped inside the filter cartridge. Plastic filter cartridges are available in many different diameters, materials, and pore sizes. A typical choice for removing, for example, remaining Pd/C particles from an alcoholic dispersion is a filter with 0.5 μm pores. Note that

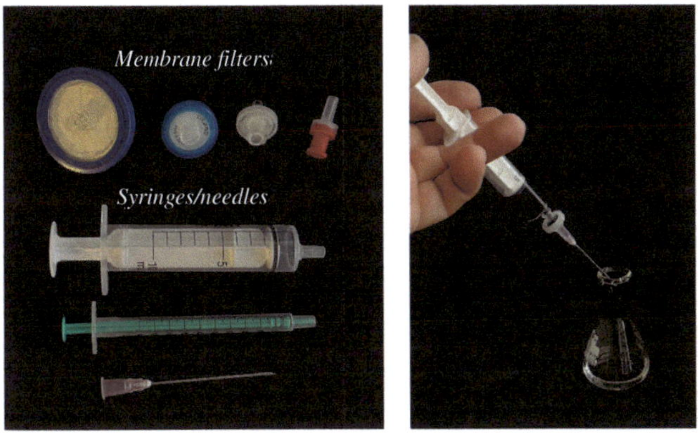

(a) (b)

Figure 4.12 (a) Plastic membrane filters and syringes used for clean-up filtrations. (b) Use of a membrane filter to clean up, for example, water or alcohol solutions. Note that special membrane materials are needed for organic solvents such as chloroform, dichloromethane, acetone, ethyl acetate, acetonitrile, and DMF, among others.

some plastic filter cartridges are incompatible with organic solvents like chloroform, dichloromethane, acetone, ethyl acetate, acetonitrile, DMF, etc.

4.4 Evaporation

4.4.1 Introduction

Evaporation is a very common operation in organic chemistry. Since reactions and purifications are usually carried out with dissolved compounds, the solvent sooner or later has to be removed to get pure compounds. Solvents used in organic chemistry are quite volatile and evaporation is usually the best way to get rid of the solvent (if the dissolved compound is not volatile). Before the 1950s, evaporations were done by distilling off the solvent using ordinary distillation setups. Since this operation was carried out quite often, there was room for time-saving improvements.

4.4.2 The Rotavapor

In the mid-1950s, the Swiss Company Büchi introduced the “rotavapor,” which is essentially a distillation apparatus fully dedicated to evaporation, preferably under reduced pressure. The distillation flask is rotated in order to speed up evaporation and avoid bumpy boiling. During operation, the flask is immersed in a water bath that can be heated. The solvent vapors are condensed in a cooling device, and the condensed liquid is collected in a large round-bottomed receiver. The cooling device uses either a liquid, usually running tap water, or a solid, e.g., ice or solid carbon dioxide (dry ice), for cooling.

Two rotavapor varieties are shown in Figure 4.13, one with water cooling and the other with dry ice cooling. It did not take long for the rotavapor to become a household item in organic chemistry labs. It is nowadays so integrated in lab works that the

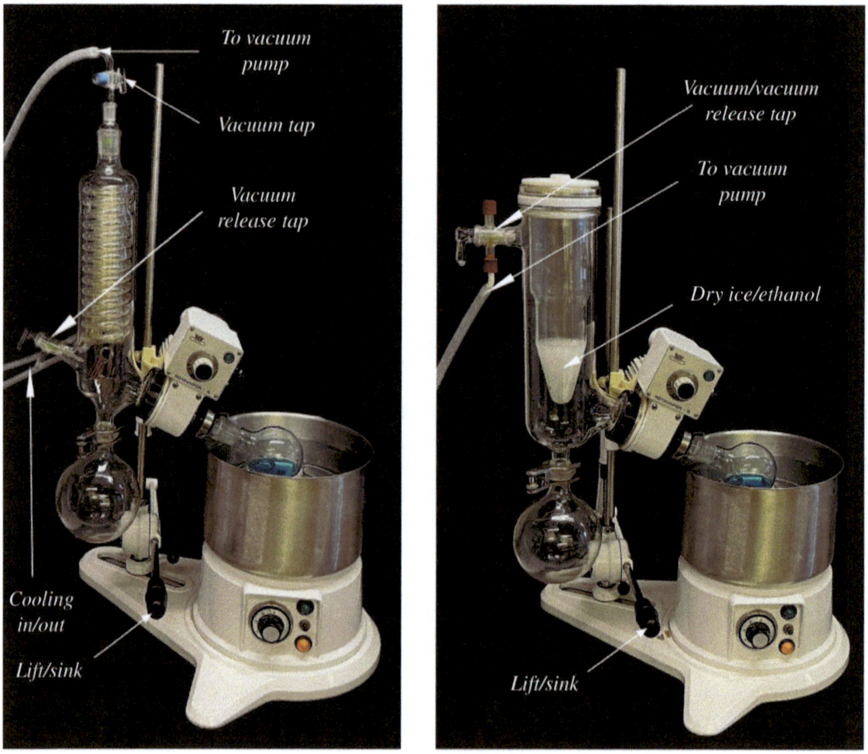

Figure 4.13 (a) Rotavapor with water cooling. (b) Rotavapor with dry ice cooling. The distilling flask (the one with the blue liquid) is rotated during the entire evaporation process, which increases the solvent surface/evaporation rate and reduces unwanted bumpy boiling.

literature phrases "was evaporated under reduced pressure" and "was concentrated *in vacuo*" simply mean that the solvent had been distilled off with a rotavapor.

Rotavapors are nowadays available in many different sizes and from many different manufacturers. A typical lab-scale rotavapor can handle solutions ranging from mL up to 2 L in volume. Recent models may have some extra items (e.g., LCD displays and a power lift) and look different than the early Büchi models shown in Figure 4.13, but functioning and handling remain essentially the same. Below is a step-by-step instruction on how to use a rotavapor, followed by a subsection with tips and tricks (Section 4.4.4) to make "rotavaping" easy and safe. Section 4.4.4 also describes three alternatives to "rotavaping" that can be considered for small-volume samples: "blowing in," "speedvacing," and "freeze-drying."

First and foremost, a warning concerning rotavapors: accidents can happen if they are not properly handled! Remember, we are dealing with evacuated glassware, and if the glass parts of the rotavapor are damaged they can implode and shatter glass pieces all around with great force. Use of safety googles is therefore, as usual in the lab, mandatory. Also, rotavapors should be located in lab surroundings where the risks of damage are minimized. Many keep them in fume hoods, since this also reduces the risk of vapors from solvent spills spreading in the lab.

4.4.3 Rotavapor Operation, Step by Step

a. Make sure that you have circulating coolant in the condenser or dry ice/ethanol in the "cold finger."
b. Set the water bath temperature to somewhere between room temperature and 40 °C.
c. Add your solution to a suitable, tare-weighed round-bottomed or pear-shaped flask. The solution should *not* fill the flask beyond half of its volume.
d. Attach the flask to the ground glass connection of the rotavapor. Use an adapter if necessary. Do not yet immerse the flask in the water bath, just leave it hanging in the air.
e. Secure the ground glass connection with a clip or similar! The most common mishap in rotavapor work is dropping the flask into the water bath at some point (for example, here). Do not yet immerse the flask in the water bath, just leave it hanging in the air.
f. Make sure that the "vacuum release tap" and the "vacuum tap" are both open.
g. Start the rotation: almost at full speed if the flask is <1 L, at half speed or less if it is larger. Immediately thereafter, start the vacuum pump and close the "vacuum release tap." The pressure should now decrease rapidly.
h. Keep a hand on the "vacuum tap" while you observe the rotating flask. If the solution bumps or foams, immediately close the "vacuum tap." The bumping or foaming should stop. Slowly reopen the "vacuum tap." Be prepared to close it again if there is more bumping or foaming.
i. When there is no more bumping or foaming the pressure has stabilized, and the flask has become cold (a sign that evaporation has started): lower the rotating flask down into the water bath.
j. Check that the vapor condensation has started, which is usually indicated by drops from the condenser/cold finger. Some highly volatile solvents do not condense here at all, but at least you should be able to see that the solution in the flask is losing volume over time.
k. Wait until there appears to be no more solvent left in the flask and the pressure is stable. The residue in the flask can be a solid or a thick syrup.
l. Turn off rotation and lift the flask out of the water bath. Turn off the heating.
m. Slowly open the "vacuum release tap" and switch the vacuum pump off (in this order!). Remove the flask from the rotavapor and weigh it.

4.4.4 Rotavapor Evaporation, Tips and Tricks

1. *Rotavapors – solvent residues*: no matter how efficient your equipment is, there will be traces of solvent remaining in the distillation flask residue. Two tricks can be used to remove the last traces of solvent: (i) redissolve the residue in a *highly* volatile solvent, like diethyl ether, dichloromethane, or chloroform, and evaporate again. Repeat if necessary. (ii) Remove the distillation flask and expose it to high vacuum for a few hours. Check the weight periodically. Both tricks can be used, in the above order.

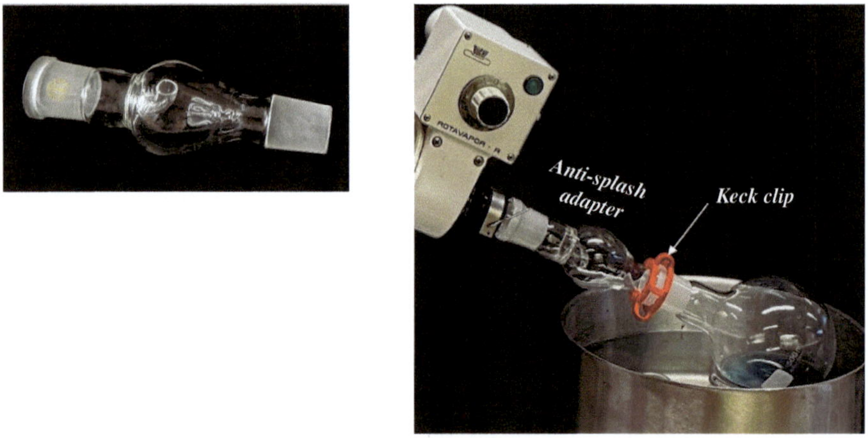

Figure 4.14 Anti-splash adapter (NS29/29). Many other designs and sizes are available.

2. *Rotavapors – bumping/splashing*: despite the constant rotation of the distillation flask, bumpy boiling can occur. If you overfill the distillation flask (more than half) you are asking for trouble, water and methanol are especially problematic. To reduce the risk of condenser contamination/substance loss, you can use an "anti-splash adapter" between the rotavapor and the distillation flask (see Figure 4.14).
3. *Rotavapors – vacuum sources*: in the early rotavapor days, water aspirators (see Section 8.2 of Chapter 8) were used to provide the vacuum. Water aspirators use tap water and provide a pressure of 10–20 mbar, which is enough to evaporate the most common organic solvents (diethyl ether, dichloromethane, chloroform, ethyl acetate, methanol, ethanol, toluene, etc.) at comfortable (<40 °C) bath temperatures.

   Although cheap and simple to use, water aspirators have nowadays largely been displaced by diaphragm pumps (see Section 8.2 of Chapter 8). A two-stage diaphragm pump gives a pressure similar to that of a water aspirator (10–20 mbar). Evaporations of solvents like water or DMF, however, require a lower pressure (1–5 mbar), which is reached only with a *three-stage* diaphragm pump. Alternatively, you can use a rotary vane oil pump (see Section 8.2 of Chapter 8) for these solvents. Such pumps easily establish a 0.1–1 mbar pressure if the rotavapor is leak-tight.

   Finally, a note about "house vacuum": larger lab complexes often have a central vacuum ("house vacuum") arrangement. House vacuum should be used with caution for rotavapor evaporations. Sudden drops or spikes in pressure can occur because many users are connected, and they can cause unwanted excessive boiling and splashing.
4. *Rotavapors – vacuum gauges*: the recommendation for rotavapor work is to always have a vacuum gauge connected, so that the working pressure can be monitored. Most modern diaphragm pumps have a built-in gauge, otherwise a simple device like a mechanical or Pirani gauge can be used (see Section 8.2 of Chapter 8). The gauge should be installed somewhere between the pump and the rotavapor.
5. *Rotavapors – cooling media*: the simplest cooling medium is tap water. If that water is not sufficiently cold or too expensive, cooling can be provided by ice or, more commonly, dry ice. In the dry ice case, some ethanol or acetone is usually added

to the dry ice (see Figure 4.13b) for a more even distribution. A third option is a central cooling medium (usually recirculating cold water with additives) which is often available in larger lab complexes. Such systems are slightly overpressurized, so extra attention should be paid here to leak-proof tubing and connections.

6. *Rotavapors – the water bath*: use only deionized or distilled water in the bath, and change contents at regular intervals. Ordinary tap water is less suitable, since it gives ugly deposits and can be corrosive.
7. *Rotavapors – servicing*: a lab rotavapor needs very little servicing. The most common service/replacement item is the gasket providing a seal between the condenser housing and the rotating glass tube (see Figure 4.15). A higher-than-normal working pressure indicates a leak, and a faulty gasket could be the reason for it. Always keep a few spare gaskets, and use the proper replacement procedure (early Büchi models required a special tool!).

   Another common service action is cleaning. You can clean the rotating glass tube and the condenser housing by partially dismantling and spraying with solvents from a spray bottle. To avoid dismantling, you can use a more spectacular method: mount a two-necked round-bottomed flask half-filled with wash solvent onto the rotavapor and apply vacuum while stoppering the second neck of the flask with your thumb. When the suction becomes painfully strong, quickly remove the thumb. The sudden pressure increase results in spraying of the rotavapor interior with solvent, which then drains into the receiver from where you can dispose it (see Figure 4.16).

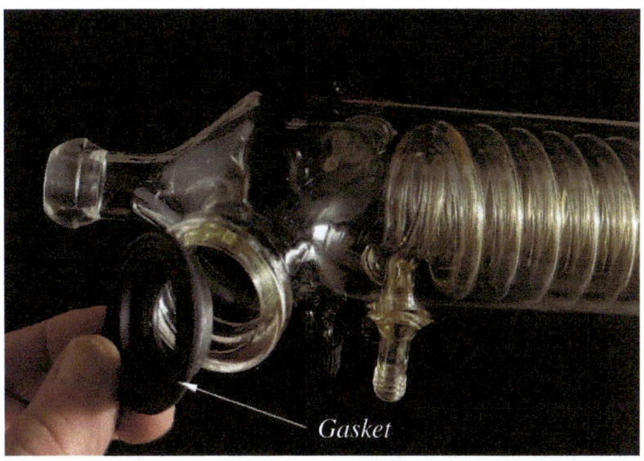

Figure 4.15 The gasket between the condenser housing and the rotating glass tube. Size and design can vary between different rotavapor models. Depicted here is the Büchi variant.

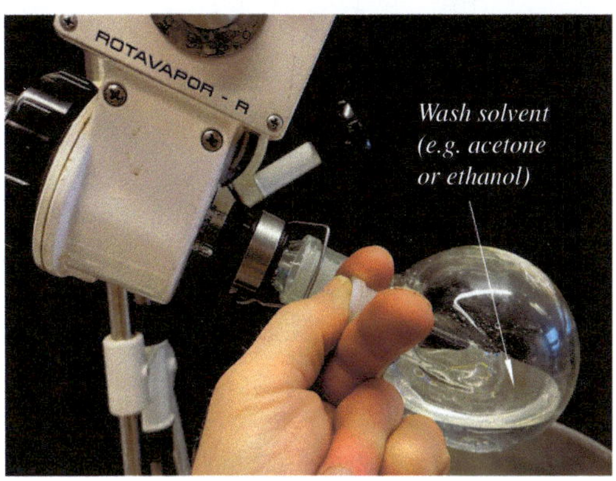

Figure 4.16 Washing trick: suction followed by sudden thumb release sprays wash solvents inside the rotavapor.

4.4.5 Alternatives to Rotavapors

There is no absolute need for a rotavapor for *all* types of evaporations. If you are dealing with only a few milliliters of a volatile solvent and only a few samples, then all you need is a fume hood and a source of compressed air or nitrogen. The gas is used to "blow away" the solvent (see Figure 4.17). A syringe or similar attached to a rubber tubing serves as a "nozzle." Adjust the flow of gas to a *very* gentle stream. Slowly direct the gas stream to an open vial with your solution, making sure that there is no splashing. When the arrangement is stable, fix the "nozzle" with a clamp, and wait until the solvent has evaporated (Figure 4.17). Warming of the vial in a water bath speeds up the process, but a room-temperature gas completes the evaporation by itself if you wait long enough. The equipment can be modified to accommodate several evaporations simultaneously.

Note that, as with "rotavaping," some solvent can remain in the residue, which can be removed by exposure to high vacuum until there is no more weight loss (overnight is usually enough).

A second alternative to the rotavapor is the speedvac, which is often used in biochemistry labs, where evaporation of multiple small (2 mL or less) samples is common. Here you put the solutions, often contained in tiny plastic test tubes, in a rotor which then centrifuges under vacuum and temperature control until the samples are "dry." The centrifugation eliminates splashing when vacuum is applied. A typical speedvac apparatus (see Figure 4.18) consists of a centrifuge enclosed in a vacuum chamber, a cold-trap to condense the vapors from the samples, and a good vacuum pump (usually of the rotary vane oil type).

A third alternative to the rotavapor is the freeze-dryer. Freeze-dryers use the fact that frozen water sublimes readily under low-pressure conditions. They are used for

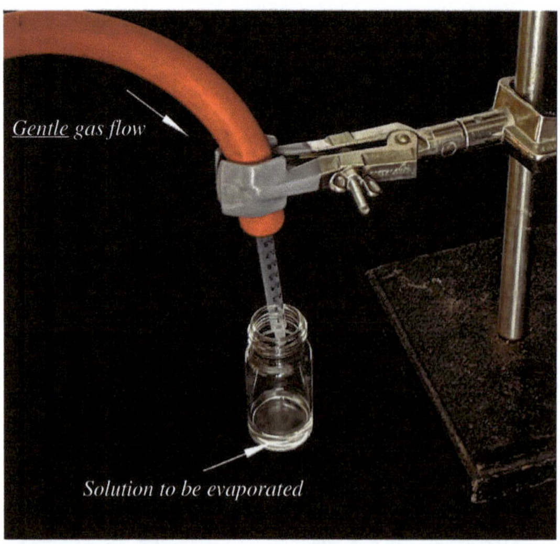

Figure 4.17 Evaporation of small amounts of solvent with the aid of a gas stream. Always in a fume hood!

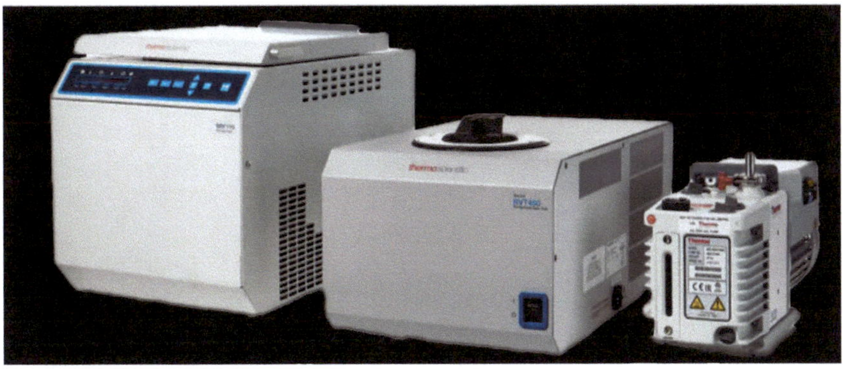

Figure 4.18 A speedvac system, from left to right: vacuum chamber with a centrifuge inside, cold-trap, and vacuum pump.

evaporation of temperature-sensitive aqueous solutions under *very* mild conditions. More details about freeze-drying can be found in Section 4.6.

4.5 Distillation

4.5.1 Introduction

Distillation is probably the most spectacular of the separation techniques. A cloudy and miscolored liquid is heated in a vessel, and the vapors are led into a cooling device where they are transformed into clear, colorless drops. The whole process is very visual! No wonder that the ancient craftsmen who could perform distillations (pharmacists, alchemists, monks, and brewers) were regarded as wizards.

In today's organic chemistry labs, distillations are not performed as often as they once were. Of course, a synthetic step that produces a distillable product still calls for distillation setup and procedure (see Figures 4.19 and 4.20), but distillations for purification of solvents or reagents have become very unusual. The main reason for this is that pure and dry reagents/solvents have become readily available from chemical vendors in almost any volume and quality. The busy chemist no longer needs to purify solvents or reagents by setting up own distillations in the lab, as was the case before.

Evaporations with a rotavapor, on the other hand, have become more common. They could be regarded as distillations, but the focus is on the residue, not the distillate as in a typical distillation. The details and tricks of rotavapor use have therefore been treated in a separate "evaporation" section (4.4.2). The present subsection will, with pictures, step-by-step instructions, and tips, demonstrate typical lab-scale distillations at normal atmospheric or moderately reduced pressure. For descriptions of other distillation techniques/procedures ("molecular" distillation, spinning band columns, high-vacuum distillation, etc.), the interested reader is referred to literature on the subject [1–5].

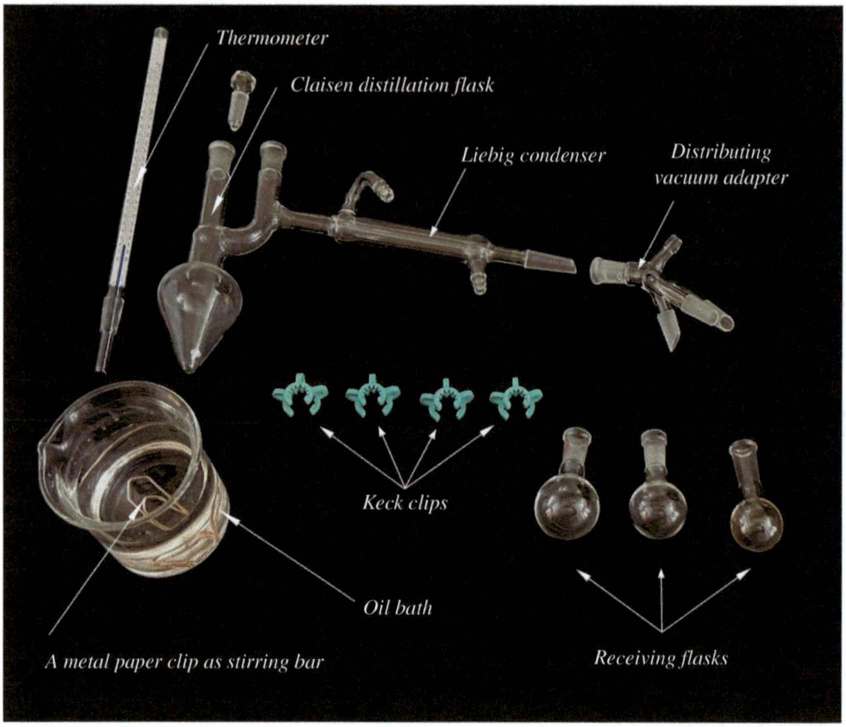

Figure 4.19 Equipment that you need for small-scale (5–25 mL) distillations. In addition to the items shown here, you need a variable heat source (e.g., a stirrer/heater, see Figure 4.20). For vacuum distillations, you also need a vacuum source (see Section 8.2 of Chapter 8) and a manometer.

4.5.2 Atmospheric Pressure Distillation

Liquids that boil at <200 °C can often be distilled at atmospheric pressure if they are reasonably stable at their boiling point. This is the simplest type of distillation.

4.5.2.1 Step-by-step Procedure, Small-scale Atmospheric Pressure Distillation

a. Set up the distillation equipment as in Figure 4.20, do not forget a stirrer magnet in the Claisen flask! Tare weigh all receiving flasks! Use a small funnel or a Pasteur pipette to add the liquid to be distilled (bluish in Figure 4.20) into the distillation flask. Note how much you have added (weight, if possible). To avoid splash-overs during distillation, do not fill the flask beyond half of its volume.

b. Start the cooling water, adjust the flow (observe at the outlet) to a slow stream.

c. Start the stirring, it should be quite vigorous. Begin heating the oil bath, let the temperature slowly increase, aim for 10–20 °C above the expected distillation boiling point. Observe the distillation flask carefully during the heat-up.

d. When the bath temperature is getting close to the boiling point of your liquid, you may see the first signs of boiling, but the real action comes when the bath temperature has increased a bit beyond the boiling point.

Figure 4.20 Typical setup for atmospheric pressure distillation (scale: 5–25 mL). Note that only one clamp is used in this small-scale example. Support (e.g., a loosely tightened clamp) at the end of the condenser is optional here, but must be used on large-scale setups.

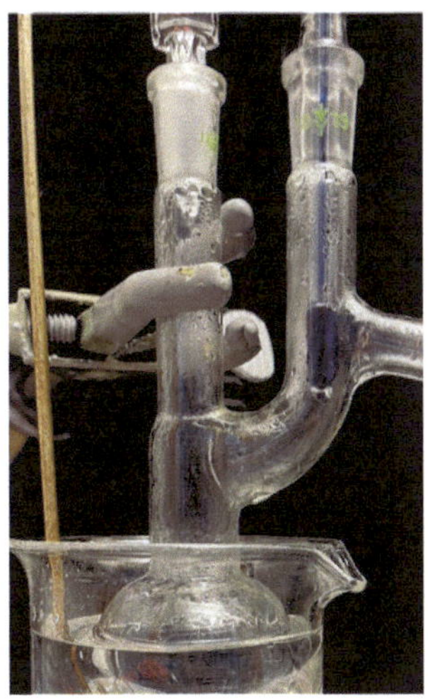

Figure 4.21 An ongoing distillation. Note the condensed drop on the thermometer bulb.

e. Once boiling has started, you will see a "vapor zone" traveling up the neck of the flask, eventually reaching the thermometer bulb. When that happens the thermometer, which has not reacted much before, suddenly displays a near-boiling-point temperature, and a condensed liquid drop appears on the thermometer bulb (see Figure 4.21). It should remain on the bulb throughout the distillation.

f. The vapor zone now quickly moves past the thermometer bulb and into the condenser, and soon you will see the first drops falling into the first receiving flask. Note the thermometer reading when that happens. After a short time, the thermometer reading should stabilize, then note the reading and change to the second receiving flask. This one will be your main flask if you are distilling a one-component liquid.

g. Observe the thermometer continuously during the main flask filling, and try to maintain a steady drop count. If it decreases, increase the bath temperature. At the end of the distillation, the vapor zone often retracts and the thermometer reading drops. Note the temperature when this happens and change to the third receiving flask.

h. Try to squeeze out the last drops from the distillation flask into the third receiving flask by increasing the bath temperature further. Note the temperature interval for the third flask, as you have done for the other two. When no more action is seen, stop the distillation by lowering the oil bath away from the distillation flask, using the lift. Turn the heat off, carefully remove the receiving flasks, and check their weights. Compare with the initial weight and calculate the distillation yield.

4.5.3 Reduced Pressure Distillation ("Vacuum Distillation")

Liquids that become unstable at their "normal" boiling point can be distilled at reduced pressure (the boiling point decreases when the pressure is lowered). This type of distillation is called vacuum distillation. The equipment is very similar to that for atmospheric pressure distillation, but in addition a vacuum source (see Section 8.2 of Chapter 8) is needed.

4.5.3.1 Step-by-step Procedure, Small-scale Reduced Pressure Distillation (Vacuum Distillation)

a. Set up the distillation equipment as in Figure 4.22, do not forget a stirrer magnet in the Claisen flask! Tare weigh all receiving flasks! Use a small funnel or a Pasteur pipette to add the liquid to be distilled (it is bluish in Figure 4.22) into the distillation flask. Note how much you have added (weight, if possible). To avoid splash-overs during distillation, do not fill the flask beyond *one-third* of its volume. This volume is less than for atmospheric pressure distillations, since the splashing/bumping risk is much higher now.

b. Start the cooling water, adjust the flow (observe at the outlet) to a slow stream.

c. Start the stirring, it should be quite vigorous. Start the vacuum pump (it should be able to provide 5–20 mbar "vacuum"). Wait until the manometer gives a pressure reading in the expected range. Be prepared for leak searching.

d. Begin heating the oil bath, let the temperature slowly increase, aim for 10–20 °C above the expected boiling point. Observe the distillation flask carefully during the heat-up.

e. When the bath temperature is getting close to the boiling point, you may see the first signs of boiling/bumping, but the real action comes when the bath temperature has increased a bit beyond that temperature.

Figure 4.22 Typical setup for reduced pressure distillation ("vacuum distillation," scale: 5–25 mL). The setup is almost identical to the one in Figure 4.20, with the differences being yellow-marked:
(i) a vacuum pump with manometer is connected to the vacuum adapter; and
(ii) two clamps are used instead of one, the right one should be loosely tightened to avoid tensions in the glassware.

f. Once boiling has started, you will see a "vapor zone" traveling up the neck of the flask, eventually reaching the thermometer bulb. When that happens the thermometer, which has not reacted much before, suddenly displays a near-boiling-point temperature, and a condensed liquid drop appears on the thermometer bulb (see Figure 4.21). It should remain on the bulb throughout the distillation.

g. The vapor zone now quickly moves past the thermometer bulb and into the condenser, and soon you will see the first drops falling into the first receiving flask. Note the thermometer and pressure readings when this happens. After a short time, change to the second receiving flask. This will be your main flask if you are distilling a one-component liquid.

h. Observe the thermometer and manometer continuously during the main flask collection, and try to maintain a stable drop count. If it decreases, increase the bath temperature. At the end of the distillation, the vapor zone often retracts and the thermometer reading drops. Note the temperature and pressure when this happens and change to the third receiving flask.

i. Try to squeeze out the last drops from the distillation flask into the third receiving flask by increasing the bath temperature further. Note the temperature and pressure intervals for the third flask, as you have done for the other two. When no more action is seen, stop the distillation by lowering the oil bath away from the distillation flask, using the lift. Turn the heat off, open the vacuum vent, turn off the pump, and carefully remove the receiving flasks and check their weight. Compare with the initial weight and calculate the distillation yield.

4.5.4 Distillation Tips/Tricks

1. *Distillation yields*: a total yield of more than 90% is unusual, regardless of the type, scale, and the liquid's initial purity. Drops of condensed liquid adhere to the apparatus walls, the heat in the distillation flask causes partial degradation into tarry nonvolatile products, and so on. It is especially difficult to achieve good yields when doing small-scale distillations.
2. *Smooth boiling*: a smooth boiling, without "bumping" or "splashing," is crucial for all distillations. In both the above examples, mechanical stirring is used to achieve this. Another way to achieve smooth boiling is to add a small amount of tiny ceramic particles ("boiling stones") to the distillation flask. When heated, these particles release air bubbles that induce boiling. Note that boiling stones are only used in atmospheric pressure distillations, they are useless in vacuum distillations. Mechanical stirring works better there. Another trick, only used for vacuum distillations, is to let a very thin glass capillary (see Figure 4.23) protrude from the outside and down into the bottom of the distillation flask. This provides a steady stream of small bubbles, which promotes boiling. If the

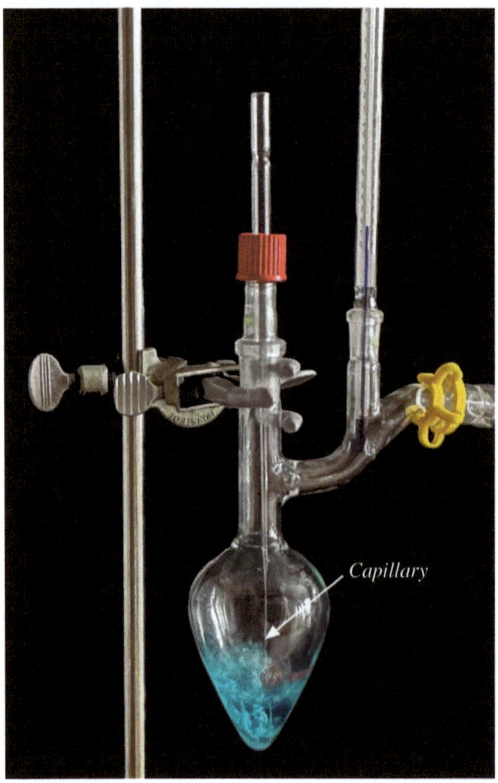

Figure 4.23 A very thin glass capillary (here drawn from the tip of a Pasteur pipette) extending to the bottom of the distillation flask provides a stream of small bubbles when the vacuum is applied. This reduces the risk of bumpy boiling.

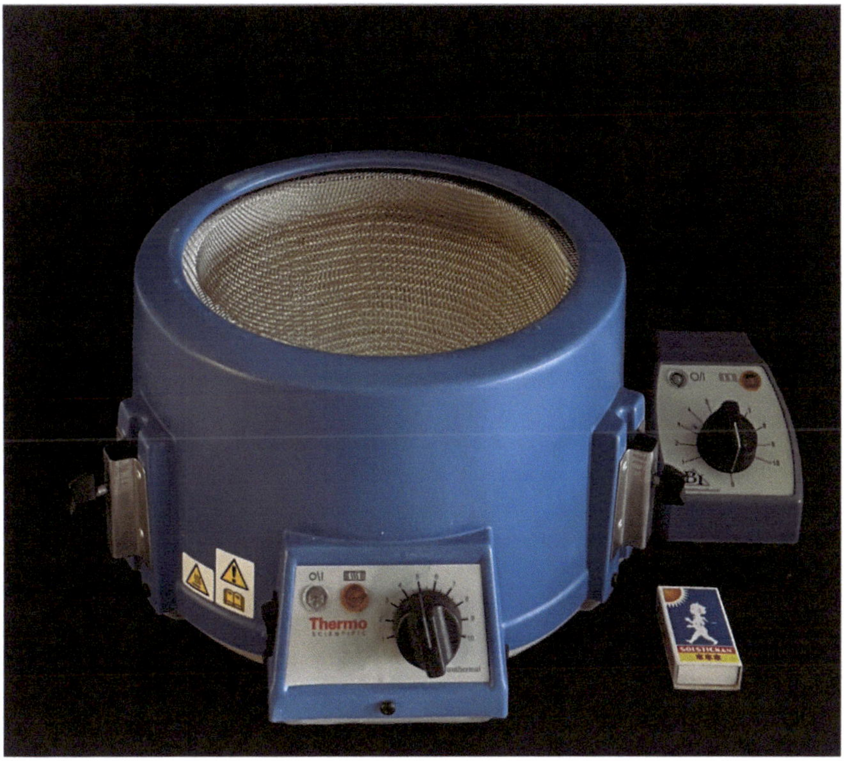

Figure 4.24 An electric heating mantle suitable for a 5 L round-bottomed distillation flask. At this scale, oil baths are no longer practical.

capillary is sufficiently thin,[1] the bubbles will be very small and the "vacuum" will not be too much affected by the leak.

3. *Heating devices*: in Figures 4.20–4.22, a hot polyethylene glycol (PEG) oil in a beaker is used for heating. This is only one of several heating options. Another is to use an electric heating mantle (Figure 4.24). This avoids the oil, but puts restrictions on the shape and size of the distilling flask. Heating mantles are primarily used for large-scale distillations (liters or more). A third option is to carefully heat the distilling flask directly with a small burner or heat gun. Obviously, this is not a very safe procedure, so use it only for small-scale, short-path distillations (milliliter or less, see Figure 4.25).
4. *Oil bath hazards*: oil baths before the 1960s were of the high-boiling petroleum type. The hot oil represented a considerable fire hazard, especially if it came into contact with water. An explosion, similar to when hot frying oil is mixed with water, could occur. For this and other reasons, the less dangerous PEG oils are used nowadays in heating baths (water baths can, of course, also be used for achieving temperatures under its boiling point). The PEG oil is fully miscible with

1 A good capillary blown from a Pasteur pipette should create bubbles in diethyl ether but not in water when blown through with the mouth.

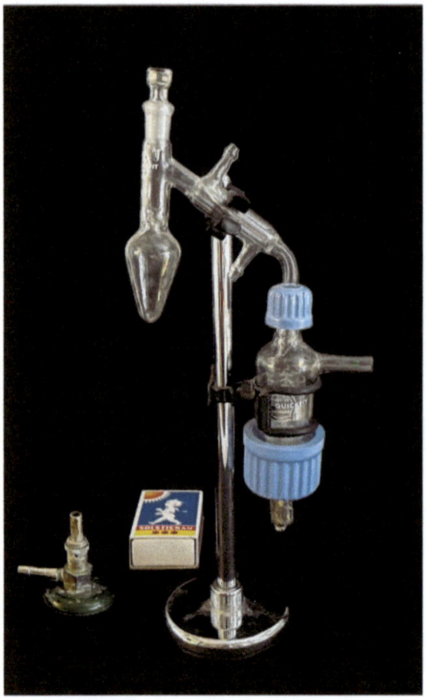

Figure 4.25 A microscale (0.5–5 mL) distillation apparatus. At this scale, heating with a burner or small heat gun is often preferred.

water and represents less of a fire hazard. However, PEG oil is hygroscopic and can, if left in an open vessel for longer periods, absorb quite a lot of water. When a water-contaminated PEG oil is heated above 100 °C, it starts to boil off the water and the bubbling can become quite irritating. To avoid water contamination, it is important to keep PEG oil in closed bottles during longer periods of storage.

5. *Oil bath stirring*: to maintain an even temperature distribution in an oil bath during heating, it should be stirred. An office paper clip (see Figure 4.19) is a better alternative than an ordinary Teflon magnet, since it gives a more gentle but still sufficient stirring.

4.6 Freeze-drying ("Lyophilization")

4.6.1 Introduction

Freeze-drying is a gentle method for evaporating aqueous solutions to dryness. It is based on the fact that frozen water sublimes readily under certain (low-)pressure conditions. When a frozen aqueous solution of a sample is kept at room temperature below a pressure of 6.1 mbar, the water starts subliming and the sample slowly "dries up," remaining frozen during the process. Sensitive compounds like proteins or nucleic acids have a better chance of surviving during this process than during lengthy room-temperature evaporations under vacuum in, for example, a rotavapor. In principle, it is possible to freeze-dry a sample with the help of only a cold-trap, some tubing, and a good vacuum pump (Figure 4.26a). That's all it takes, but the capacity is somewhat limited. Therefore, a dedicated all-in-one laboratory apparatus relying on the same principles (see Figure 4.26b) is usually preferred, despite the higher cost. Large and even more sophisticated freeze-dryers are used in the pharmaceutical and food industries. A speedvac is a close relative to the freeze-dryer, but here the samples are not necessarily frozen. Read more about speedvac equipment in Section 4.4.

4.6.2 Freeze-drying, Step by Step

Freeze-drying can be done with many types of vessels and sizes. Below is a typical procedure for lab glass flasks (10–500 mL):

- **a.** Transfer the aqueous solution to be freeze-dried into a large-enough round-bottomed or pear-shaped flask (tare-weighed!). The flask should be less than half full. One-third full or less is a good choice.
- **b.** Freeze the sample solution with dry ice or liquid nitrogen (you can also put it in the freezer, but it takes longer). If possible, rotate the flask for faster freezing and larger surface area. Wait until the entire solution is deep-frozen. Faint cracking sounds indicate that the freezing process is finished.
- **c.** Check that the freeze-dryer vacuum is below 1 mbar. Attach the flask with the frozen solution to the freeze-dryer's rubber extrusion and open the extrusion's vacuum valve *very slowly*. The pressure will rise momentarily, then drop again. Check that it stabilizes, preferably below 1 mbar.
- **d.** After several hours (overnight is best) check your sample, it should now be a dry solid. Close the extrusion's vacuum valve *very slowly* (the solid can be "fluffy" and spread all over if the air rushes in too quickly).
- **e.** Remove your sample, check the weight, and subtract the tare weight.

(a)

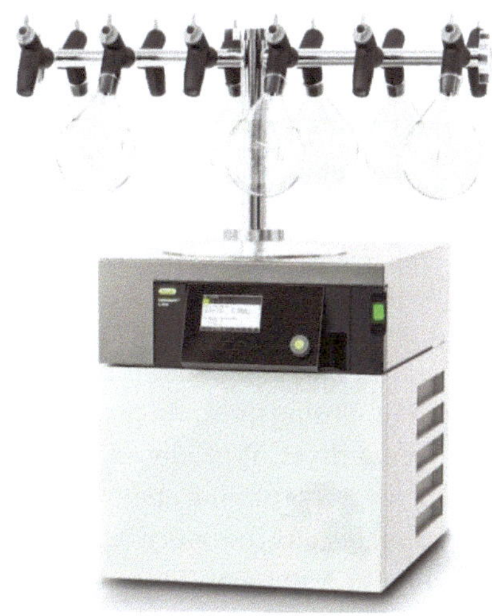

(b)

Figure 4.26 (a) A very simple freeze-dryer: a vacuum pump maintains a sufficiently low pressure that the water in the frozen sample sublimes into the cold-trap. (b) An all-in-one benchtop freeze-dryer with attached flasks containing frozen samples. The "box" has manometer and temperature displays in front. A vacuum pump and a large cold-trap are hidden inside the box.

4.6.3 Tips, Tricks, and More Information

1. *Sample pretreatment for successful freeze-drying*: in many cases, the sample solutions contain not only water but also other solvents or additives. For example, HPLC fractions can contain considerable amounts of acetonitrile or methanol, and sometimes also small amounts of formic or trifluoroacetic acid. Most water-miscible solvents "disturb" the freeze-drying process, often to the extent that the sample melts and freeze-drying turns into a (very slow) room-temperature evaporation of a liquid. Also, many volatile solvents other than water do not condense well in the cold-trap/condenser, instead they make their way into the pump oil, leading to decreased pump performance. Even worse, volatile acid vapors can cause permanent damage.

   A trick to avoid problems during freeze-drying is to pre-evaporate "risky" sample solutions with a rotavapor that has a pump efficient enough to quickly remove unwanted solvents/additives at room temperature. Ideally, the evaporation should also remove some of the water. This reduces the freeze-drying time, and also improves the "compactness" of the residue after freeze-drying. Last but not the least, pre-evaporation reduces the risk of pump oil contamination (see also below under "servicing").
2. *Servicing*: Like most laboratory apparatus, freeze-dryers need regular servicing. The most frequent service item is the cold-trap/condenser which sooner or later has collected so much ice that the cooling becomes less efficient and there is a risk of clogging. Defrosting is then needed and, just like with household freezers, it means shutting everything down, thawing/removing ice, and restarting. Typically, defrosting is needed several times per month on a constantly running lab freeze-dryer where large sample volumes (>50 mL) are frequently dried. To determine when defrosting is needed, many freeze-dryers have "windows" through which the condenser coils can be observed. Problems with maintaining a stable vacuum can also be an indication that it is defrosting time.

   The second most frequent service item is the vacuum pump oil (the pump is usually of the rotary vane/silicone oil type). The oil degrades slowly over time, and even the best cold-traps cannot prevent small amounts of water or other contaminants from accumulating in the oil. So, just as with car combustion engines, the oil needs to be replaced regularly. This means shutting everything down, emptying out all the old from the pump by opening the pump's drain plug, closing the drain plug, refilling with fresh oil, and restarting the pump and freeze-dryer. Typically, oil changes are done every second or third month on a constantly running and frequently used lab freeze-dryer.
3. *Nonaqueous freeze-drying*: freeze-drying is an efficient procedure for evaporation of water, but it can also be used to evaporate certain other solvents, with similar benefits. One example is benzene (melting point: 5.5 °C). Solutions of lipophilic compounds in this solvent freeze-dry very well and often provide nice and fluffy powders, whereas ordinary evaporation gives thick syrups. Dimethyl sulfoxide (DMSO, melting point: 19 °C) and acetic acid (melting point: 16 °C) solutions can also be freeze-dried.

4.7 Crystallization/Recrystallization

4.7.1 Introduction

Crystallization is a purification method that, just like filtration and distillation, was used long before there were any chemistry labs. Salt, sugar, and many other everyday materials are products of crystallization. In organic chemistry, crystallization (often called recrystallization) is used to remove impurities from a crystalline solid. The solid is dissolved in a solvent until a relatively saturated solution is obtained. Then the conditions are changed (e.g., by cooling) so that the solubility is decreased (i.e., the solution becomes "oversaturated"). This will induce crystallization, leaving the impurities in solution. The crystals are then filtered off and dried. Examples of a successful crystallization are seen in Figures 4.27 and 4.28.

Figure 4.27 Crystals of 4-nitrophenol.

Figure 4.28 Crystals of 4-nitrophenol (enlarged).

There are several ways to carry out crystallization, depending on how the solubility is manipulated. The most common way is to use a single solvent, make a nearly saturated hot solution, and cool it slowly to induce crystallization (the solubility of a compound usually decreases with temperature). This variant is called "crystallization from a single solvent."

Another way is to use two solvents, one where the compound is very soluble ("good solvent") and another where it is sparingly soluble ("bad solvent"). The solubility of the compound can then be varied by both composition changes and temperature changes. Addition of more "bad solvent" to a nearly saturated solution will induce crystallization, and so will cooling. This variant is called "crystallization from two solvents."

A third way to crystallize a compound is by evaporation of a solution until it becomes "oversaturated." It is seldom used in organic chemistry but very often in industry, e.g., to prepare salt from seawater or pure sucrose from beet or cane.

The two first variants are described below. Note that these procedures should be conducted in a well-ventilated hood, with appropriate fire-protection devices within reach whenever flammable solvents are used.

4.7.2 Crystallization from a Single Solvent, Step by Step, Scale: 0.1–5 g of Solid, Figure 4.29a–g

Figure 4.29a Weigh out the impure solid (0.1–5 g) in a suitable Erlenmeyer flask. The flask should well contain the amount of solvent you may need. Save a few milligrams of the solid as "seed" and reference material. Put the flask on a magnetic stirrer hotplate. Add a stirring magnet.

Figure 4.29b Add a suitable solvent (for selecting solvent, see Section 4.7.4). Use a volume that is deemed somewhat *insufficient* to dissolve all material. Cap the Erlenmeyer flask with an inverted beaker or similar loose-fit device.

Figure 4.29c Stir and heat the solution to gentle boiling. The solid should now at least partially dissolve. Add extra amounts of solvent and continue boiling. Repeat until all solid has dissolved (small amounts of insoluble impurities could remain, see Section 4.7.4).

Figure 4.29d Remove the flask from the hotplate, fish out the stirring magnet, and let the solution cool down slowly to room temperature.

Figure 4.29e After some time, crystals should begin to form. If crystals do not form spontaneously, you can seed ("nucleate") with a few milligrams of the original solid. Another way to induce crystallization is to gently scrape the walls of the flask with a glass rod or a spatula.

Figure 4.29f The crystallization process is finished when no more crystal growth is seen. Wait at least an hour. Overnight is best. Often the crystals form a lattice or a "cake" that should be broken up by stirring or shaking before the next step, filtration.

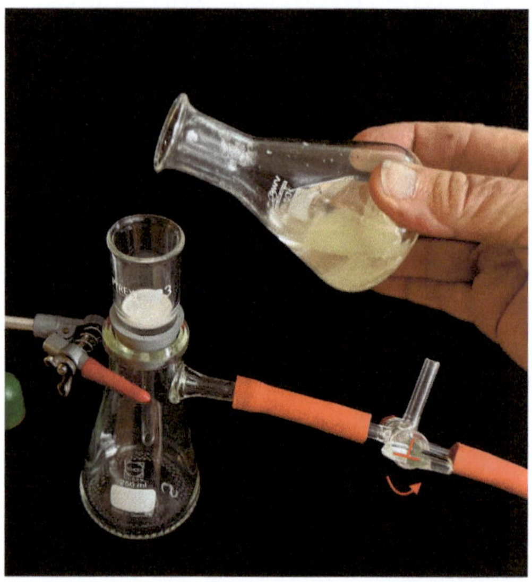

Figure 4.29g Filter the crystal "slurry" by suction filtration (for details, see Section 4.3). Transfer the crystals to a tare-weighed container. Air-dry them or use a vacuum desiccator. Note the weight, and compare with the weight of the starting solid. Note that you can evaporate the filtrate ("mother liquor") to dryness and repeat the crystallization process to give a second batch of crystals.

4.7.3 Crystallization from Two Solvents, Step by Step, Scale: 0.1–5g of Solid, Figure 4.30a–f

Figure 4.30a As with the one-solvent procedure, mix the solid with solvent, this time of the "good" type and a rather small amount.

Figure 4.30b Heat and stir until all solid has dissolved.

Figure 4.30c Start adding "bad solvent" while stirring continuously and heating gently.

Figure 4.30d Continue adding "bad solvent" until there is a slight persistent turbidity.

Figure 4.30e Add drops of "good solvent" until the turbidity disappears and the solution becomes clear again. Usually, a few drops are sufficient!

Figure 4.30f Remove the flask from the hotplate, fish out the stirring magnet, and let the solution cool down slowly to room temperature. Seed if necessary. When no more crystals form, continue with filtration and washing as in the one-solvent case (use a cold mixture of the two solvents for washing).

4.7.4 Tips and Tricks

4.7.4.1 Large-scale Crystallizations

When solvent amounts exceed 500 mL, you should replace the simple Erlenmeyer flask setup shown above with a more elaborate setup, consisting of a round-bottomed flask fitted with a reflux condenser and stirring magnet. Heating is with either a water bath, an oil bath, or a heating mantle.

4.7.4.2 Finding a Suitable Solvent

When recommendations for solvent/solvent mixtures etc. are not available, you must figure out suitable conditions yourself by doing some trial experiments. Heat a small (10–50 mg) amount of impure solid in a 4 mL test tube with a solvent (1 mL). If it dissolves, cool and seed with a tiny speck of the solid. If you see some nice crystal growth (check with a magnifier!), then the solvent is suitable. If there are no crystals or you cannot dissolve the solid, try another solvent, and so on. Figure 4.31 shows the equipment that you need.

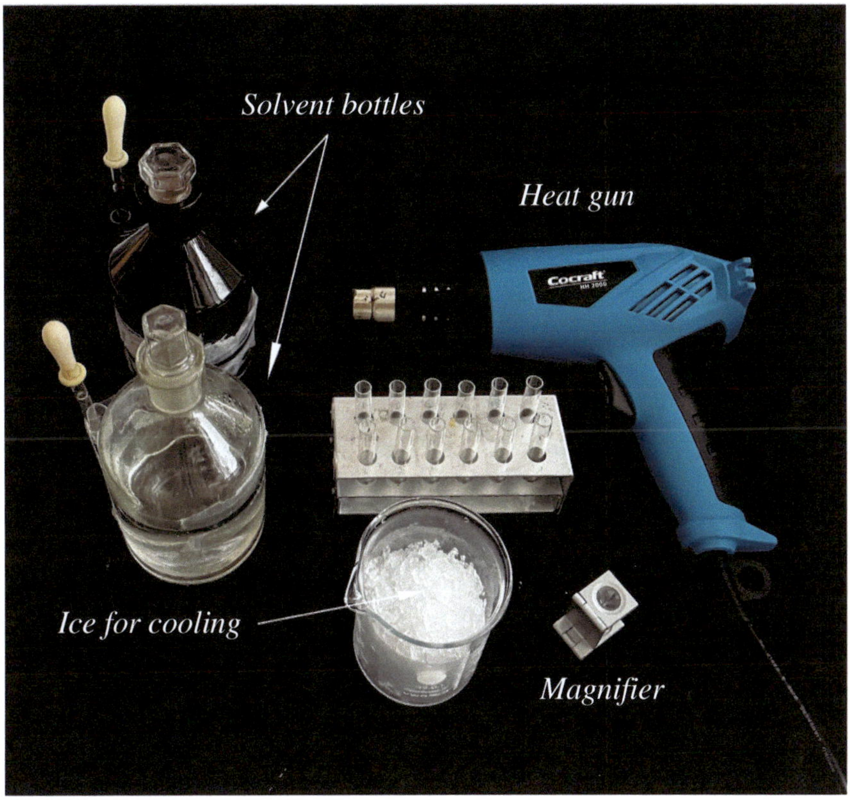

Figure 4.31 Equipment for finding a suitable solvent/solvent mixture.

Some common single solvents, in the approximate order of decreasing polarity, are given below:

- Water
- Methanol
- Ethanol
- Acetone
- Dichloromethane
- Chloroform
- Ethyl acetate
- Pentane
- Heptane
- Petroleum ether

Some common solvent pairs, in the approximate order of decreasing polarity, are given below:

- Water–methanol
- Water–ethanol

- Methanol–dichloromethane
- Ethyl acetate–pentane
- Dichloromethane–pentane
- Chloroform–pentane

Some "rules of thumb" when choosing a solvent are as follows:

a. Nonpolar/hydrophobic compounds often crystallize well from polar solvents, especially if the impurities are polar.
b. Polar compounds often crystallize well from nonpolar solvents, especially if the impurities are nonpolar.
c. Chloroform, dichloromethane, and ethyl acetate are popular "good solvents" in two-solvent crystallizations.
d. Petroleum ether, pentane, and hexane are popular "bad solvents" in two-solvent crystallizations.

4.7.4.3 If Oil or Emulsion Appears Instead of Crystals

In crystallization of solids with low melting points, you often get an oil or an emulsion instead of crystals. Then there is no purification effect at all. The chances to get crystals increase if you lower the temperature where the "saturation" point is reached. Therefore, it is often better to go from room temperature to fridge or freezer temperature rather than from boiling to room temperature. However, remember that filtration of crystals formed in the cold should also be done in the cold and as quickly as possible.

Warning: flammable solvents in fridges or freezers are a potential hazard if the flasks leak vapors. Stopper them safely!

4.7.4.4 Decantation or Hot Filtration

It sometimes happens that minor impurities refuse to dissolve even when heating with an otherwise suitable solvent. You should remove these solids before crystallization begins. If they sediment rapidly, a simple decantation of the hot solution may be sufficient (see Figure 4.32a). In other cases, you need to filter the solution while hot ("hot filtration"). This is a tricky process, since crystallization often begins already during filtration if the filter is not heated. There are different ways to heat a filter. On a small (0.1–5 g) laboratory scale, you can do it by using a "fluted" filter paper (hand-folded from a circular filter or purchased, Figure 4.32b) in a funnel which is heated from underneath by vapors rising up from boiling solvent in the filtration flask (Figure 4.32c). Caution: in all hot filtrations with volatile and/or flammable solvents, it is *extra* important that you consider the safety aspects. Always operate in a fume hood, and have fire-fighting devices at hand.

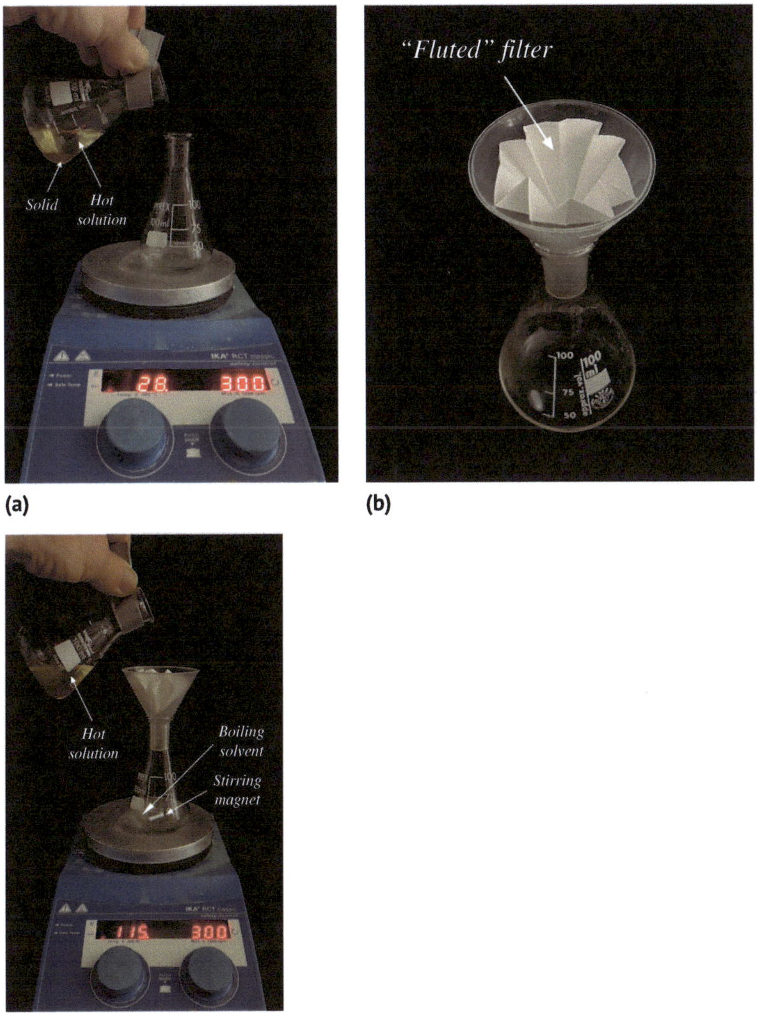

(c)

Figure 4.32 (a) Decantation, (b) fluted filter paper, and (c) hot filtration.

4.7.4.5 Seeding

When you seed a clear solution with a small amount of crystals, three things can happen:

a. *The seed crystal dissolves*: then your solution is not saturated. Cool, concentrate, or, if two solvents are used, add more "bad solvent."

b. *The seed crystal induces crystallization*: this is your lucky day.

c. *The seed crystal neither induces crystallization nor dissolves*: this could be called "crystallization inhibition," which can have several causes. One is that the impurities in the solution are much less soluble than the main compound and therefore "precipitate" onto the seed crystal, making it useless as a seed. When this happens, it is better to use another solvent system, perhaps at the other end of the polarity scale.

If seed crystals are not available, crystallization in a solution can be really difficult to induce, even with suitable solvents and concentrations. The compound stubbornly remains dissolved or precipitates as an oil or amorph, even in cases where it is known that the compound *can* crystallize. You should then do several different attempts to get the first seed crystals. One trick is to leave a sample of the compound for days in a concentrated form in a glass vessel, or let a solution of the compound slowly evaporate in a glass vessel. Frequent scratching of the vessel glass walls with a glass rod or similar can help, and both room temperature and fridge temperatures could be tried.

Another trick is to purify the compound rigorously with some chromatographic technique, and then do crystallization trials again. Pure compounds usually crystallize easily and more spontaneously than impure ones.

However, there are numerous examples of compounds that, despite being reasonably pure, refuse to crystallize until one day, months or years later, small crystals suddenly appear, and after that further crystallization is easy. Getting sluggish compounds to crystallize is something of an art. The most skilled in the art can get even large proteins to crystallize. But that is a different story, told in other books.

4.8 Column Chromatography

4.8.1 Introduction

Column chromatography was first described in 1905 by the Russian scientist Michail Tsvet [6]. He used a calcium carbonate column percolated with petroleum ether/ethanol for the separation of plant pigments. Since then, many types of column materials and solvents have been tried and evaluated. The earliest column materials (carbonates, aluminum oxide, and silica gel) separated mixtures according to their adsorption properties toward polar materials. More recently developed materials separate also according to other properties, like molecular weight (size-exclusion chromatography) or specific binding to certain molecules (affinity chromatography).

In synthetic organic chemistry, silica gel column chromatography is an indispensable tool for preparative purification of reaction mixtures. Most new compounds reported today have been purified by preparative column chromatography, often as the final step in a "workup" procedure.

A detailed procedure for "open-column" manual chromatography is given below (gravity mode on silica gel). Some chemists use the "flash" modification (see Section 4.8.3) or commercial, prepacked columns in combination with a pump for their chromatographic purifications. The solvent selection process and general workflow are rather similar in all these three variants.

During the last decades, sophisticated computer-controlled column chromatography systems (Figure 4.33) have become available [7]. These systems integrate a prepacked column, a pump, a UV detector, and a fraction collector in a single unit. They can save considerable amounts of time for the frequent user, but are more resource-demanding in terms of purchase and maintenance. Some extra user training is also needed. A detailed description of computer-controlled column chromatography systems is unfortunately beyond the scope of this book.

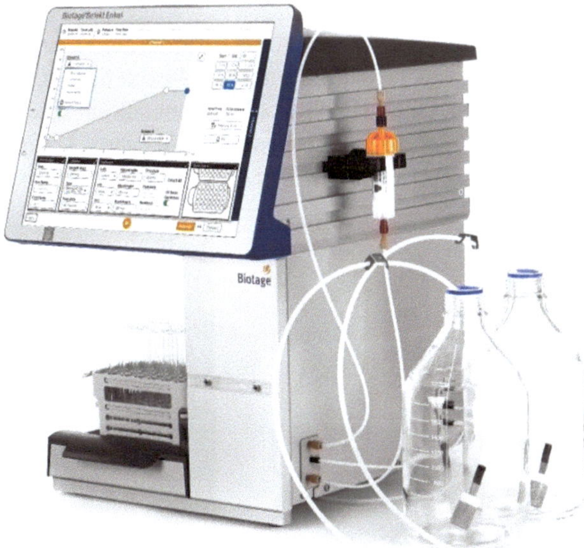

Figure 4.33 An example of an integrated column chromatography system. Solvent reservoirs, a prepacked column, pumps, a detector (usually UV-based), and a fraction collector are included. The system is computer-controlled.

4.8.2 Manual Open Column Chromatography on Silica Gel, Step by Step, Figures 4.34–4.38

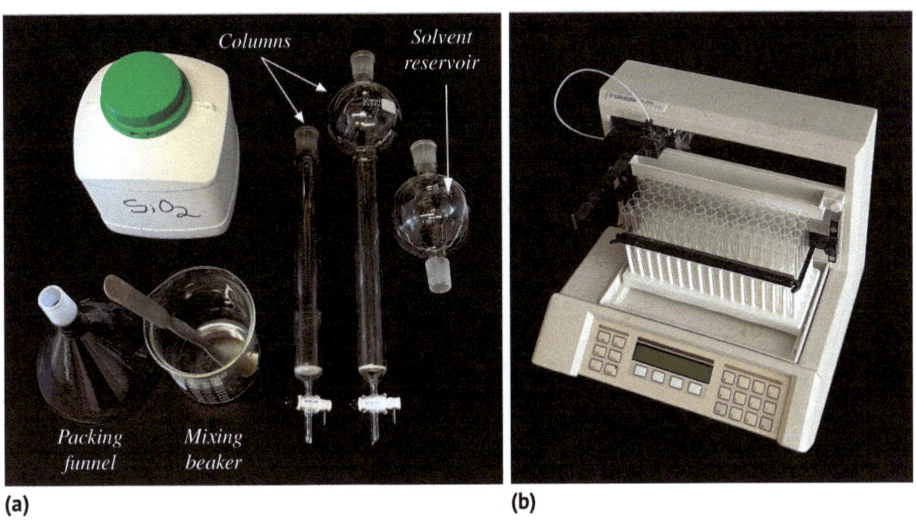

Figure 4.34 Equipment needed for manual open column chromatography: (a) the basics and (b) a simple fraction collector (optional).

4.8.2.1 Solvent Choice and Packing

Caution: all column packing and running operations should be performed in a well-ventilated hood! Inhalation of solvent vapors and silica dust is a health hazard.

First, with the aid of silica gel TLC plates, try out a suitable solvent/solvent mixture for your sample. For details about how to do this, see Section 3.2 of Chapter 3. The goal is to find conditions where your major compounds have R_f values between 0.3 and 0.5.

Figure 4.35a Weigh out the amount of dry silica gel needed into a large beaker. A rule of thumb is that the weight of the silica should be 50–100 times the weight of the sample mixture to be purified.

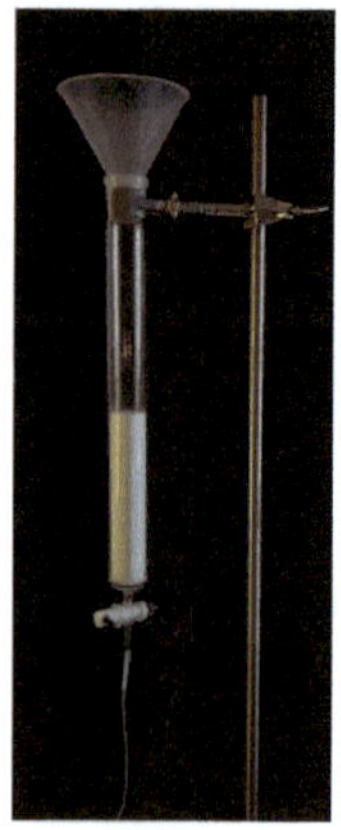

Figure 4.35b Select a suitable glass column for the job. To check, pour the dry silica gel into the column with the aid of a packing funnel. The silica gel should fill up the column to a maximum of three-fourths of its length.

Figure 4.35c Pour the dry silica gel back into the beaker.

Figure 4.35d Add enough of the chosen solvent to the silica gel to make a thin slurry after stirring. No bubbles!

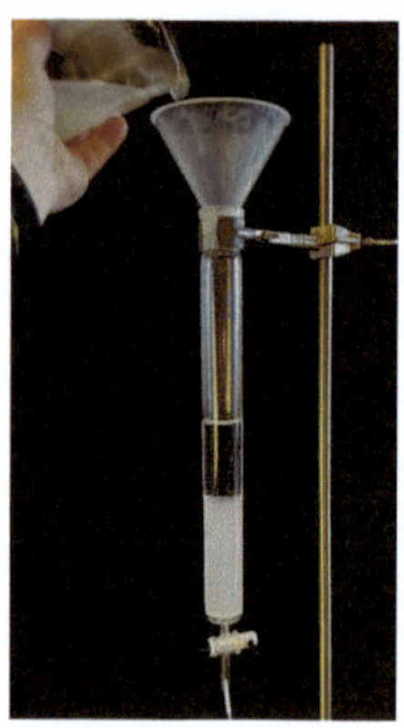

Figure 4.35e Attach a packing funnel to the column, and close the tap. Fill the bottom one-fifth of the column with solvent. Then pour in the silica gel slurry (all at once if possible), and let the material settle for a minute or two.

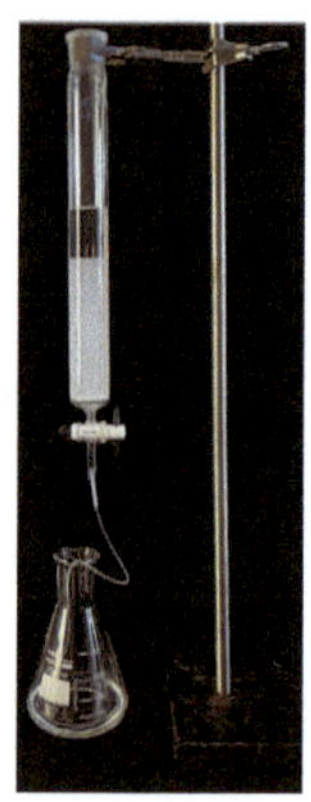

Figure 4.35f Open the tap and let out the solvent until only a few centimeters are left above the silica gel surface. Remove the packing funnel. If needed, attach tubing to the column outlet. The column is now ready for use.

4.8.2.2 Sample Application and Elution

First, make a 10–30% solution of the sample (note the dry weight!) to be purified, preferably in the same solvent that the column was packed in. Set aside a few microliters of the sample solution as a reference.

Figure 4.36a Lower the column solvent level down to just a few millimeters above the silica gel surface by opening/closing the lower tap.

Figure 4.36b Continue to lower the liquid level to the point where the silica gel has almost dried up ("draining," i.e., no "meniscus" visible).

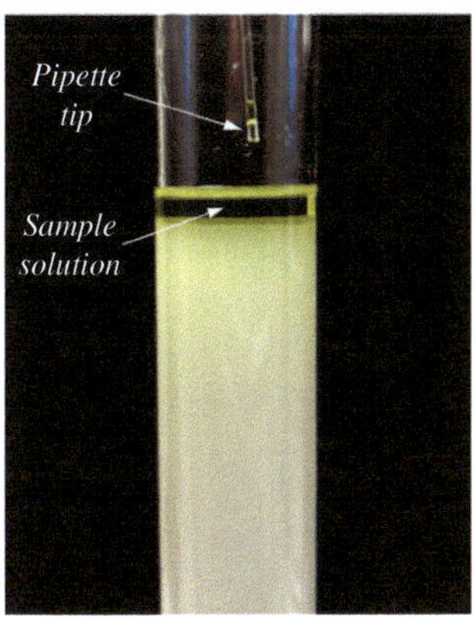

Figure 4.36c Carefully "layer" the sample solution on top of the silica gel with the aid of a Pasteur pipette or similar. From the pipette, let the solution slowly sift down the inner wall of the column, and circle the pipette for even distribution. When all sample solution has been added, "drain" again to the point where the silica gel has almost dried up, then use a fresh pipette to fill up with a small amount of pure solvent.

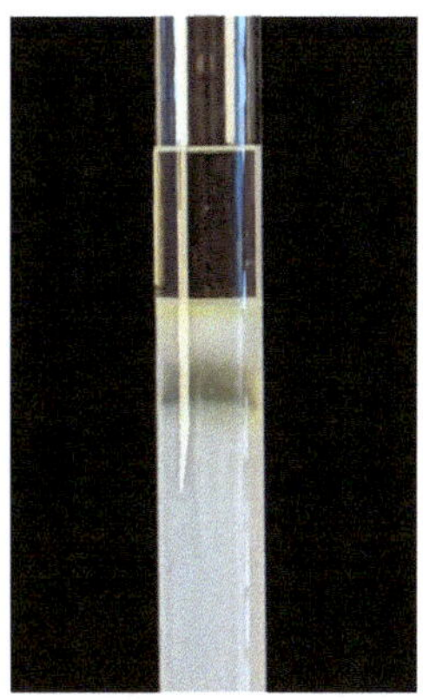

Figure 4.36d Repeat the "drain" and "fill-up" procedure a few times, so that the sample solution moves a bit down into the column packing material.

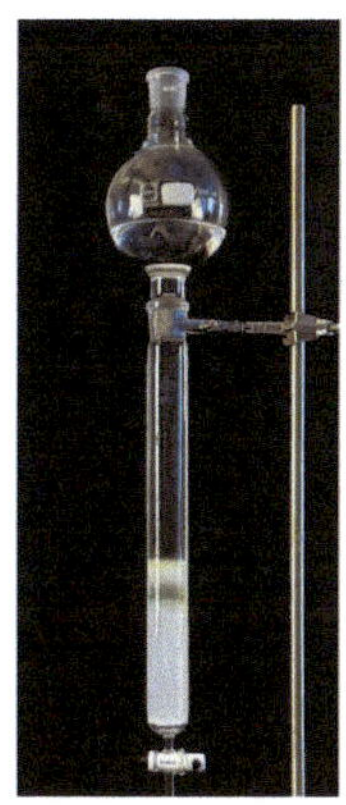

Figure 4.36e Fill up the column with solvent. Be very careful in the beginning, so that you do not stir up the silica gel. Attach a solvent reservoir on top of the column and fill it with fresh solvent. Start the elution by opening the lower tap. Collect fractions manually in a test-tube stand or use a fraction collector. Analyze the eluate continuously by doing spot tests for each fraction (see Figure 4.37). When you conclude that all your compounds have eluted from the column, stop the elution by closing the lower tap.

4.8.2.3 Fraction Checking and Pooling

Spot tests on TLC plates are a convenient way to quickly analyze the column eluate. Take aliquots from the fractions with a capillary and spot onto an ordinary silica gel TLC plate. No elution of the plate is needed! Just visualize the spots with a suitable method, e.g., UV absorbance. Figure 4.37 shows an example of what a spot test can look like.

Select the fractions where the spot test indicates presence of material, and again spot aliquots onto a *wide* TLC plate. For reference, also spot the original mixture first or last on the same plate. Elute the plate with a suitable solvent, e.g., the same that you used for column elution. Visualize the spots. An example TLC plate is shown in Figure 4.38.

Based on the TLC analysis, pool appropriate fractions in tare-weighed round-bottomed flasks and evaporate each flask with a rotavapor.

Evacuate (< 5 mbar) the flasks for a period long enough to remove the last solvent residues almost completely.

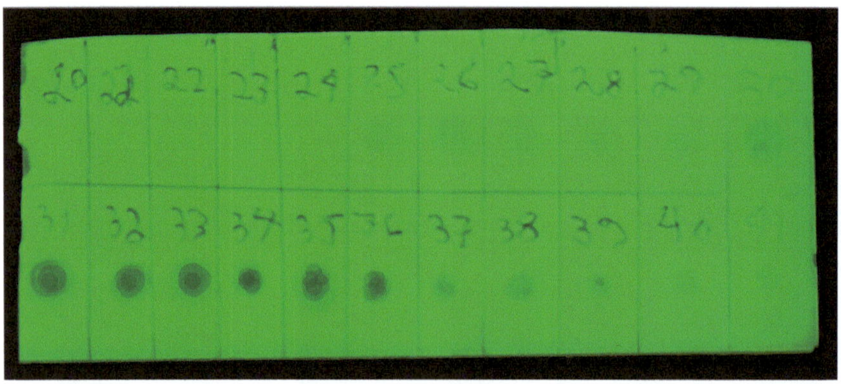

Figure 4.37 Example spot test of eluate fractions (20–41, UV visualization).

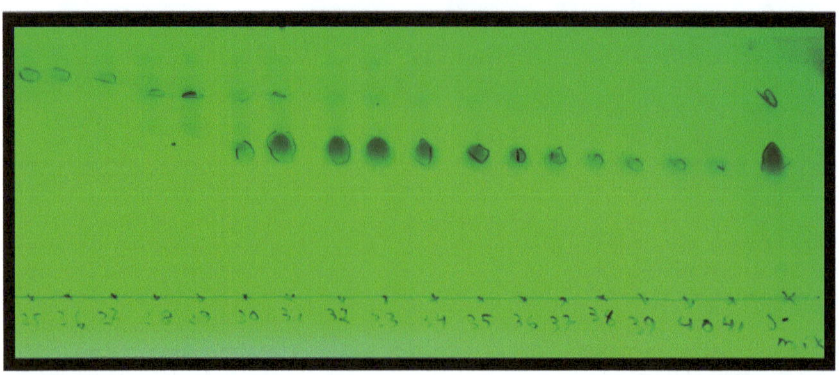

Figure 4.38 TLC analysis of the fractions (25–41) that, according to the initial "spot test" shown in Figure 4.33, contained material (UV visualization). The "S-mix" to the right is the original mixture.

Check the weights of the flasks and subtract the tare weights to get the dry weights of the contents. Compare the sum of these weights with the weight of the original mixture.

4.8.3 Tips, Tricks, and More Information

1. *Sample application*: if the sample does not dissolve well in the chosen elution solvent, you can use another solvent, preferably with a lower polarity than the elution solvent. Chloroform or dichloromethane are common choices.
   Some chemists spread a thin layer of fine sand on top of the silica gel column before sample application to prevent unnecessary stir-up of the silica gel surface during the application process. Another application trick is to mix the application solution with a small amount of silica gel, partially evaporate, and apply the semisolid material on top of the column.
2. *Gradient elution*: a slightly more time-consuming (but finally often rewarding) trick is to pack the column and start the elution with a solvent where the compounds elute very slowly (i.e., low R_f values according to TLC). After some time, the solvent is stepwise changed toward higher polarity. This is called gradient elution. If done correctly, the separation of the compounds in the mixture becomes more distinct. Gradient elution also reduces the risk that all compounds elute together with the solvent front, which can happen if you have selected a too "fast" solvent mixture.
3. *Evaporation*: with syrupy compounds, the evaporation of solvent from the fractions may not be complete. Small amounts of residual solvent can remain even after overnight evacuation. Repeated co-evaporations with a very volatile solvent (e.g., dichloromethane or diethyl ether) can improve the situation.
4. "Flash" *chromatography*: you can make a column elution much faster by applying pressure on top of the column, this modification is called "flash chromatography" [8]. A pressure-regulating head (see Figure 4.39) is attached to the column and then connected to a compressed gas source. You must take care not

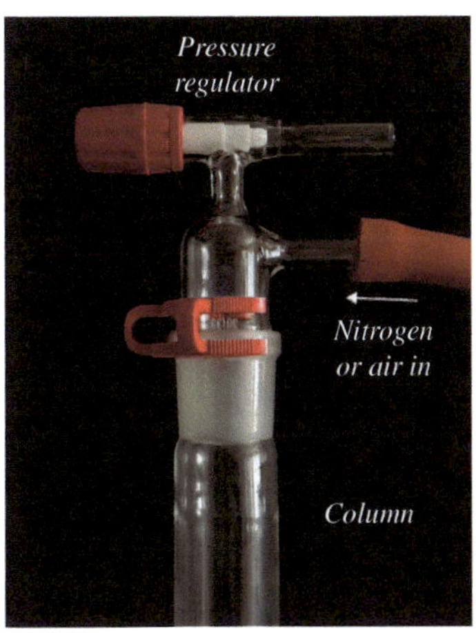

Figure 4.39 Pressure-regulating head for flash chromatography.

to apply too much pressure, since this can cause solvent splashing or other mishaps. Carried out correctly, column chromatography in the "flash" mode can be completed in less than 30 minutes (column chromatography in the ordinary "gravity mode" usually takes several hours). Note that in flash chromatography, it is more important that the silica gel particles have a narrow size distribution, otherwise the "bands" with the different compounds may broaden too much.

A milder way of applying pressure to a column is to extend it upward with a large solvent-filled reservoir like the one shown in Figure 4.36e. The increase in hydrostatic pressure on the column is proportional to the distance from the reservoir solvent surface to the silica column surface. If the reservoir can hold a large volume, that distance varies only marginally, especially during the first part of the elution. Water towers use the same principle.

5. *Solid-phase extraction*: this is a type of "on–off chromatography" that is used to "extract" water-soluble, moderately lipophilic compounds from an aqueous solution full of salts, buffers, and other very polar compounds. A small column or "cartridge" filled with reversed-phase silica material is used (Figure 4.40). When you pass an aqueous solution through the column, the lipophilic compounds are adsorbed (i.e., "extracted"), while the polar compounds pass through. After an aqueous "wash," elution with a less polar solvent (alcohols and acetonitrile are most commonly used) "desorbs" the compound from the column.

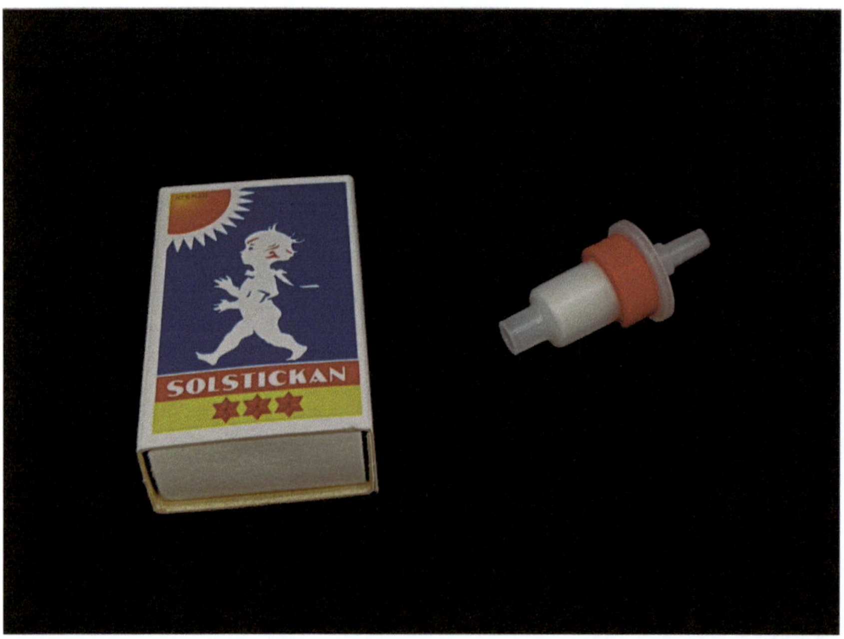

Figure 4.40 A miniature solid-phase extraction cartridge filled with 0.33 g of C-18 reversed-phase silica. The ends fit to plastic syringes and plungers. The matchbox is a size reference. Larger cartridges are also available, but when the cartridge size increases above grams of solid material, it could be cheaper to purchase the solid material separately and use ordinary glass columns.

Cartridges with other types of solid materials (ion exchange, straight-phase silica, etc.) are available and can be used in a similar way for the extraction of compounds based on other properties (charged, polar, etc.).

References

1 Leonard, J., B. Lygo, and G. Procter, *Advanced Practical Organic Chemistry*, 3rd Ed. 2013, Boca Raton, FL: CRC Press. 356 pages.

2 Vogel, A.I., et al., *Vogel's Textbook of Practical Organic Chemistry*, 5th Ed. 1989, Essex, England: Longman Scientific & Technical and John Wiley & Sons. 1514 pages.

3 Górak, A. and E. Sorensen, *Distillation: Fundamentals and Principles*. 2014, London, UK: Academic Press. 450 pages.

4 Krell, E., *Handbook of Laboratory Distillation*, 2nd Ed. 1982, Berlin: VEB Deutscher Verlag der Wissenschaften.

5 Loewenthal, H.J.E. and E. Zass, *A Guide for the Perplexed Organic Experimentalist*. 1978, London, UK: Heyden & Sons.

6 Tswett, M., Adsorptionsanalyse und chromatographische methode. Anwendung auf die chemie des chlorophylls. Berichte der Deutschen Botanischen Gesellschaft, 1905. 24: pp. 384–393.

7 Kasprowiak, A., F. Cazier-Dennin, and P.-E. Danjou, Flash chromatography system: a practical tool for demonstrating the influence of column characteristics on chromatographic resolution. J Chem Educ, 2020. 97: pp. 1145–1150.

8 Roge, A.B., et al., Brief review on: Flash chromatography. Int J Pharm Sci Res, 2011. 2(8): pp. 1930–1937.

5

Reaction Product Characterization

5.1 General Considerations

After workup/purification, the desired reaction product should be present in a pure form. To verify that this is indeed the case, the product must be *characterized*: both its purity and identity must be determined. In the early days of organic chemistry, this was done with a few robust and simple-to-understand analytical methods: melting point, elemental analysis, and optical rotation. It was also possible to estimate the molecular weight by measuring the melting point depression or boiling point elevation.

Since then, modern spectroscopic methods (especially NMR spectroscopy and mass spectrometry, MS) have revolutionized analytical organic chemistry. NMR spectra provide both purity and identity information, and MS spectra provide complementary identity information. Detailed NMR and MS data are therefore nowadays included for each new compound reported in the scientific literature.

Despite their importance, NMR and MS methods are discussed only briefly in this chapter, and the focus is on sample preparation. Chapter 3 of this book contains some more details about common LC–MS instruments and their use.

A full coverage of NMR and MS analytical methods and the theory behind them would go far beyond the scope of this book (remember, it is a *practical* guide). Many good books on these two subjects exist, some examples can be found in the literature at the end of this chapter [1–16]. Much more information is certainly available via internet searches.

Other techniques, like infrared (IR) spectroscopy, UV spectroscopy, and HPLC, are also used for characterization and the "classics," optical rotation and melting point, are also still used when needed. The basics of these two latter analytical methods are covered in this chapter. Coverage of IR spectroscopy, UV spectroscopy, and HPLC would, just like with NMR and MS, go beyond the scope of this book. The interested reader is, again, referred to the literature on these subjects, some sources can be found in [1–6, 17–19].

5.2 NMR Spectroscopy (Sample Preparation)

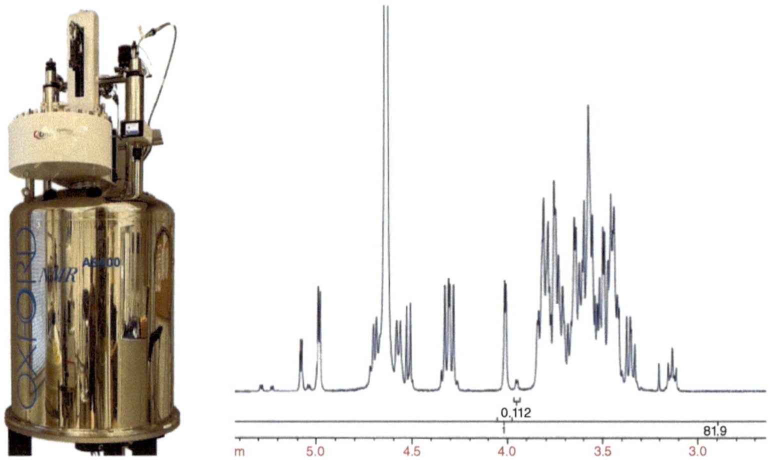

(a) (b)

Figure 5.1 (a) A superconducting NMR spectrometer. (b) An H-NMR spectrum.

5.2.1 Introduction

An NMR spectrometer (Figure 5.1a) records the radio signals that atomic nuclei in an organic molecule (most often H and C) absorb while they are exposed to a strong magnetic field. The recorded signals ("peaks") are displayed as an "NMR spectrum" of the molecule. A typical example of a H-NMR spectrum is shown in Figure 5.1b. The absorbed radio signal intensity is along the *y*-axis, while the "ppm" value proportional to the radio signal frequency is along the *x*-axis.

In H-NMR spectroscopy, the radio signal frequency is characteristic of the type of H atom that gave rise to it, e.g., —C—H protons differ from $—CH_2$ protons that differ from $—CH_2—O$ protons.

For most organic chemists, NMR spectroscopy is the single most powerful technique for structural determination. It is, however, relatively insensitive, requiring 1–20 mg of sample for a good result, with C-NMR being even more demanding than H-NMR. Luckily, the analysis is nondestructive, so the entire sample solution can be recovered.

5.2.2 Preparing an NMR Sample ($CDCl_3$ Solvent)

If you have pure crystals or liquids and you want to use $CDCl_3$ as a solvent, just weigh out or estimate 1–10 mg (5–20 mg for C-NMR) of your material and dissolve it in 0.5–1 mL of solvent. Generally speaking, the sample solution should now be filtered, for example, through a glass wool-packed Pasteur pipette (as in Figure 5.2) or a through a small PTFE filter disk (as in Figure 5.4 in Section 5.3). However, filtration can be skipped if no particles are expected or can be seen. Finally, transfer the solution to an NMR tube (see Figure 5.2). The sample is now ready for analysis.

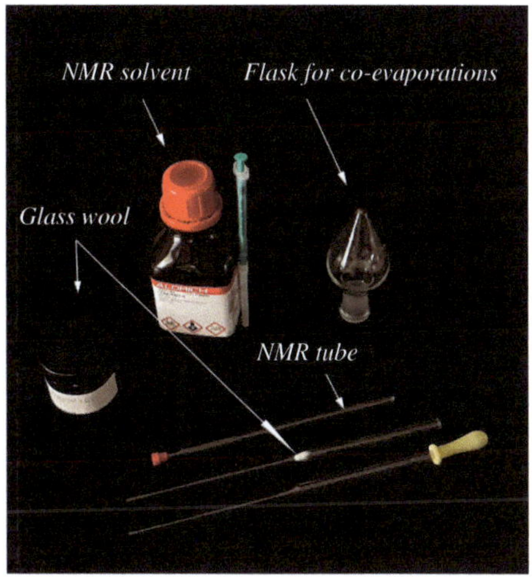

Figure 5.2 Equipment for NMR sample preparation. A pear-shaped (or round-bottomed) flask is used for co-evaporations and a glass wool pipette can be used to filter the sample solution if necessary.

5.2.3 Preparing an NMR Sample, Tips and Tricks

When you have syrupy compounds (from, for example, a silica gel column), you can expect residual amounts of chromatography solvents after evaporation. They can remain in the syrup even after long exposures to high vacuum, and their signals can obscure sample signals and make the spectrum look unnecessarily messy. You can reduce residual solvent signals by co-evaporating the sample once or twice with an undeuterated version of the NMR solvent before dissolving it in the deuterated version of the solvent. If your budget allows, you can even do a final co-evaporation with the deuterated solvent.

If you dissolve your sample in D_2O, especially if it contains OH or NH groups, you should first co-evaporate (or Freeze-dry) twice with an ordinary quality D_2O, then finally dissolve in a highest possible D_2O quality (best provided in small sealed glass ampoules). The co-evaporation will convert all eventual OH or NH groups to OD and ND groups, which simplifies the H-NMR spectrum. Also, the (unwanted) HOD signal is greatly reduced. To further reduce the HOD signal, you could "desiccate" the NMR tubes by keeping them under high vacuum until immediately before use. Alternatively, you can rinse them with ordinary quality D_2O immediately before use. Both procedures minimize the amount of adsorbed H_2O on the glass surfaces. Similar considerations apply to solutions in deuterated methanol.

5.3 Mass Spectrometry (Sample Preparation)

5.3.1 Introduction

A mass spectrometer instrument (Figure 5.3a) records the mass of ions produced from molecules when they are electrically excited, and is a very useful characterization tool in organic chemistry. There are many variants depending on the ionization method (EI, ES, FAB, MALDI, etc.), each with its own advantages and limitations. Most variants provide molecular weight information (from the "molecular ion"),

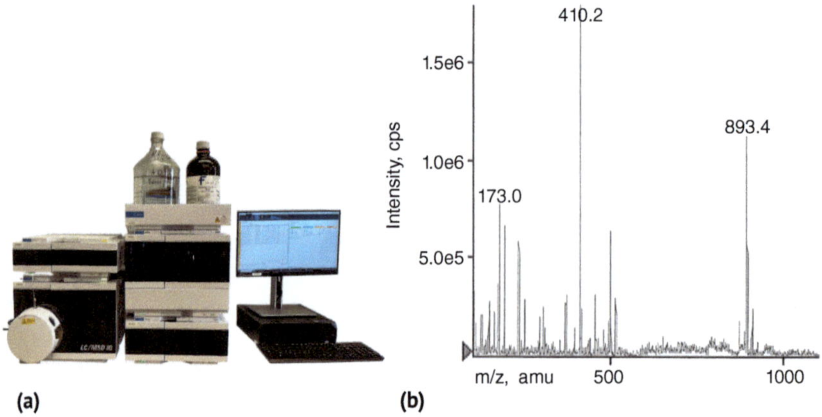

Figure 5.3 (a) An LC–MS instrument. (b) A typical mass spectrum.

sometimes even with a degree of accuracy that enables determination of a molecular sum formula. MS can also display fragment ions, providing more detailed structural information. In peptide chemistry, for example, fragment ion patterns can reveal the complete amino acid sequence!

Mass spectra are typically displayed as shown in Figure 5.3b, where signal intensity, which is proportional to relative ion abundance, is along the *y*-axis and the m/z value is along the *x*-axis. When the ion charge (z) is equal to 1, the *x*-axis value is equal to the ion mass in daltons (Da).

In preparative organic chemistry, HPLC in combination with electrospray MS ("LC–MS") has become very popular for molecular weight determinations during reaction monitoring and final characterizations. For a description of a typical LC–MS instrument, see Chapter 3. Here, only LC–MS sample preparation is discussed.

5.3.2 Preparing an LC–MS Sample

Modern LC–MS systems are *very* sensitive, requiring very tiny amounts of material to produce good spectra. Typically, microliters of 1–2 mg/mL solutions are needed. Most instruments nowadays use automatic sample injectors, which means that HPLC vials (100–1000 μL volume, see Figure 5.4) containing sample solutions are placed in a rack, from which an autoinjector withdraws a suitable sample volume (usually 1–10 μL) from each vial. The sample solvent can vary depending on which stationary phase and elution solvent the HPLC part of the instrument uses. Chemical similarity between elution and sample solvent is a good thing. Importantly, many labs require that the sample solutions be filtered before injection, e.g., through a small 0.5 μm filter disk (Figure 5.4) or through a glass wool plug inside a Pasteur pipette (Figure 5.2 of Section 5.2).

5.3.3 Preparing an LC–MS Sample, Tips and Tricks

A very important thing to remember when preparing LC–MS samples: the higher the purity, the better. MS detectors are extremely sensitive to contaminants that can

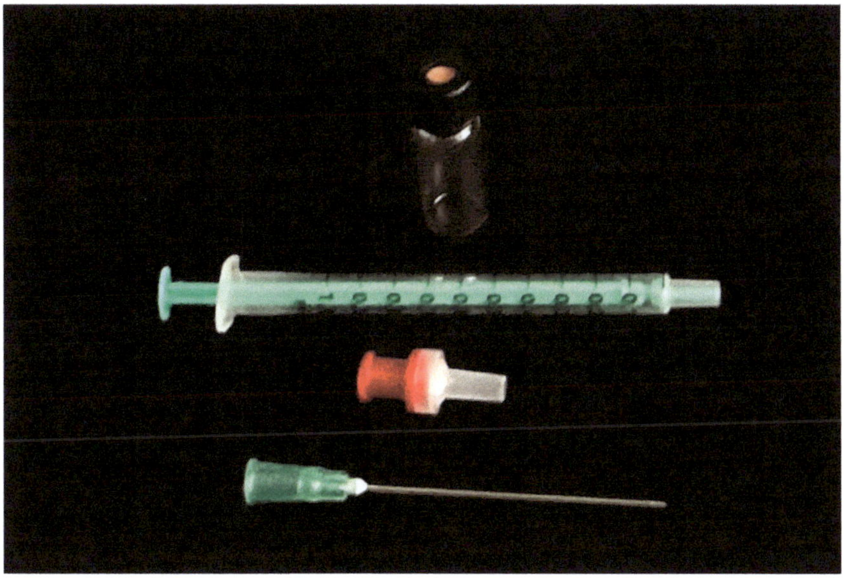

Figure 5.4 Equipment for LC–MS sample preparation. The HPLC vial shown has a volume of approximately 1 mL, but there are same-size vials with a 200 μL insert. The filter disk should be of a solvent-resistant type.

easily produce ions, e.g., salts from a reaction mixture. The LC column normally adsorbs such contaminants, but they will sooner or later exit the column and enter the MS detector. When many impure samples have been injected, the salt ions can begin to dominate each spectrum, dwarfing the analyte signals. Salt ions can remain in the MS detector ion source for a long time, constantly producing their typical signals in all spectra. Typical positive ions from salts are triethylammonium (m/z : 102), diisopropylethylammonium (m/z : 130), or pyridinium (m/z : 80).

When salt ions become a problem, a major operation, including instrument shutdown/ion source disassembly/cleaning, is needed to remove all salt residues.

When LC–MS is used for reaction monitoring, salts are often abundant in the reaction mixtures! Before analyzing salt-rich reaction mixtures, you should therefore do some sample preprocessing. In the case of ammonium-type salts, a small acidic "mini-workup" is a good routine if your product is not affected by the acid. See also Chapter 3 (Reaction Monitoring).

5.4 Optical Rotation

5.4.1 Introduction

Chiral compounds rotate polarized light, and their "specific rotation" value $\left([\alpha]\right)$ is a physical property that should be included in scientific reports of novel chiral compounds. Instruments for the determination of optical rotation are called polarimeters, and most academic chemistry labs have one. Measurements are done by putting

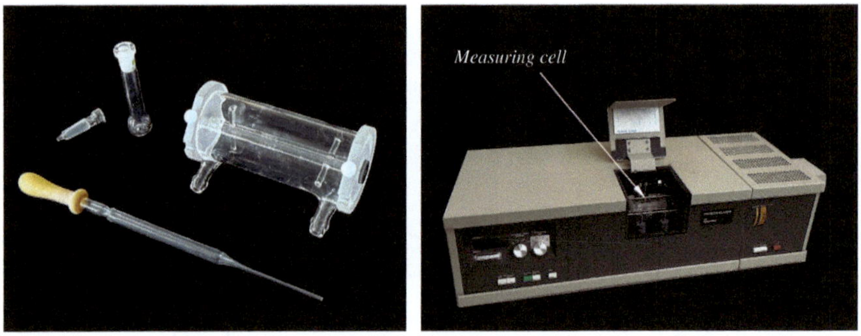

(a) (b)

Figure 5.5 Equipment for optical rotation measurement: (a) a 2 mL volumetric flask and a 1 mL/1 dm length measuring cell; and (b) a polarimeter.

a sample solution into the polarimeter's measuring cell (Figure 5.5a), inserting the cell into the instrument (Figure 5.5b), and noting the observed rotation value (in degrees). The process is nondestructive and the sample solution can be recovered.

5.4.2 Sample Preparation and Optical Rotation Measurement, Step by Step (1 mL cell)

a. Weigh out 5–10 mg of sample directly into a 2 mL volumetric flask (Figure 5.5a). Note the weight as accurately as possible.
b. With a pipette, slowly add solvent (water, alcohols, or chloroform are most commonly used) up to the line mark on the neck of the volumetric flask.
c. Shake well and check that all solids are dissolved. You now have a 2.5–5 mg/mL sample solution ready for measurement.
d. Turn on the instrument, and select, if possible, the wavelength. Most polarimeters use the "sodium D-lines" (589 nm) as the default wavelength. Let the instrument warm up for a few minutes.
e. Note the actual temperature in the room, or adjust the polarimeter's eventual thermostat setting to the desired temperature.
f. Press the zero button, and check that the instrument displays a zero value (and an acceptable signal strength).
g. Insert the *empty* measuring cell (Figure 5.5a) into the instrument and close the lid. When the instrument displays a stable value and an acceptable signal strength, press the zero button again and check that the instrument displays a zero value. You have now zeroed the instrument against the measuring cell (which could have a slight rotation power on its own).
h. With a pipette, carefully transfer the volumetric flask solution into the measuring cell, using enough to fill up. Close the cell's two openings with the provided stoppers.

i. Look through the cell from one end, the solution in the "tube" should be transparent with no bubbles or particles. If bubbles obstruct the view, tap the cell repeatedly against the desk, and check again. Emptying and refilling is sometimes necessary to get rid of all bubbles.
j. Insert the measuring cell into the instrument, and close the lid. When the instrument displays a stable value, note it down as the "first" α reading, including its sign (+ or −).
k. Remove the cell, check that the instrument displays a stable (near-zero!) value, insert the cell again, and wait for a stable display. Note down the value as the "second" α reading. It should be similar to the first.
l. Calculate the average value from the first and second readings, and consider it the "measured" α value.
m. Use the formula in Figure 5.6 to calculate the specific rotation $[\alpha]$ of your compound from the measured α value.
n. Recover the sample solution from the cell. Rinse and dry the cell carefully.

$$[\alpha]_{\lambda}^{T} = \frac{\alpha}{l \times c}$$

Figure 5.6 Calculating the specific rotation $[\alpha]$ from the measured α value. T is the measurement temperature (in °C), λ is the measurement wavelength (in nm), and l is the measuring cell length (in dm). The length is usually 1 dm and engraved into the cell glass, c denotes sample concentration (in g/mL).

5.4.3 Tips and Tricks

5.4.3.1 Confusing Mixture of Units

The units used in the specific rotation calculation do not appear fully logical and consistent, so note the following: if you have a cell length of 1 dm, it "disappears" from the calculation. You then simply divide the measured α value with the concentration in g/mL. You can also use mg/mL, but then you have to multiply the result by 1000.

Another source of confusion is how specific rotation values are reported in the literature. It can look like this: $[\alpha]_D = +23°$ ($c = 0.5$, water, 25 °C). The "D" tells you that the measurement used light with a wavelength of 589 nm ("sodium D-lines"). Water was the solvent, and "$c = 0.5$" tells you about the concentration used, here the unit is g/100 mL! This unit was common in the early days of optical rotation measurements and it is still used here, for compatibility with older reports.

5.4.3.2 Different Sample Amount, Different Measurement Precision

If you use a 1 mL instead of a 2 mL volumetric flask, you will need less sample (3–5 mg instead of 5–10 mg). However, the 1 mL measuring cell must be completely filled, so you cannot spill a drop of your solution. You also lose precision in weighing when you reduce sample quantity.

On the other hand, if you increase the sample quantity to 25–50 mg and the solution volume to 10 mL, you can use a larger (10 mL) measuring cell. It has a wider opening that allows more light through, giving a more stable and reliable instrument reading. You also gain precision in weighing when you increase sample quantity.

5.5 Melting Point

5.5.1 Introduction

The melting point is a characteristic property of a crystalline compound: it is the temperature at which the liquid and solid forms of the compound are in equilibrium. Melting points are usually determined with a melting point apparatus where you can slowly increase the temperature and observe at which temperature the compound "melts." In practice, an exact melting point is difficult to pinpoint, so a narrow temperature interval is instead reported.

In organic chemistry, the melting point has been used for a long time as a purity criterion/identity check. Nowadays, modern methods (NMR spectroscopy, MS, IR spectroscopy, chromatography, etc.) are preferred over melting points for identity checks. However, melting points are still used in, for example, the pharmaceutical industry as a simple purity criterion, since minor impurities lower the melting point significantly. In scientific publications, the melting point/melting range should be included in the description of a novel crystalline compound.

Figure 5.7a,b show the equipment needed for melting point determinations. A step-by-step procedure follows.

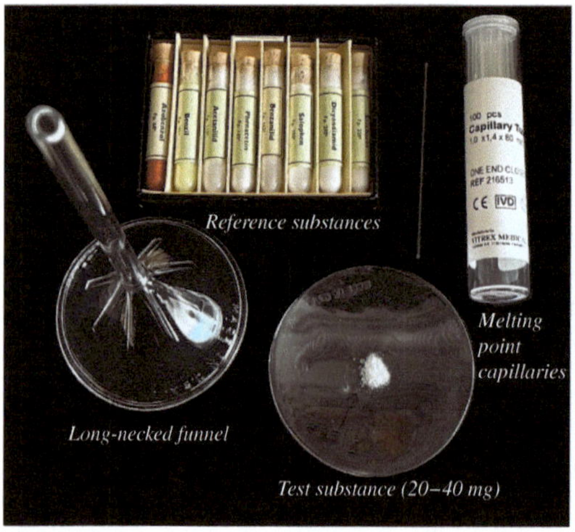

(a)

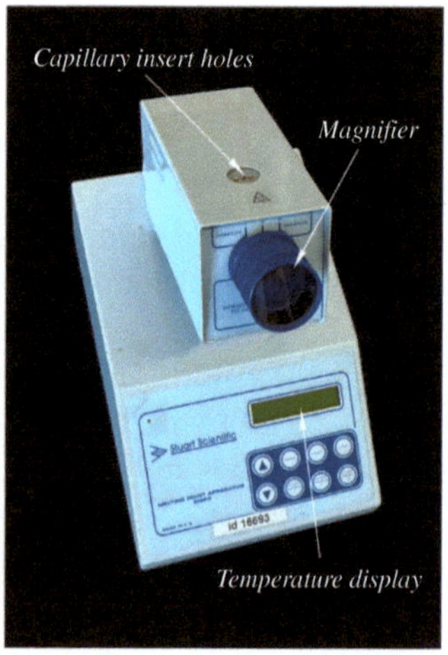

(b)

Figure 5.7 (a) Melting point capillaries, test substance on a watch-glass or similar, small glass funnel with a long pipe (or a long glass tube), and reference substances. (b) Capillary-type melting point apparatus (many other models exist with the same functions).

5.5.2 Melting Point Determination with a "Capillary-type" Melting Point Apparatus, Step by Step (Figure 5.8a–g)

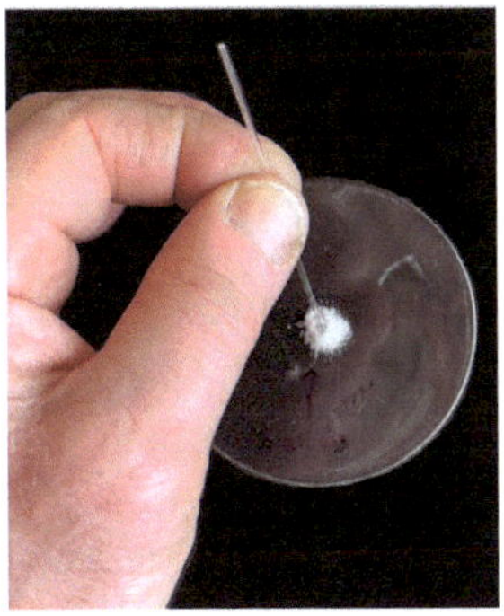

Figure 5.8a You need at least 20 mg of crystals. Press a melting point capillary's open end down into your crystal "pile" until 3–5 mm of the capillary's upper end is filled.

Figure 5.8b Turn the capillary's open end up and try to move the crystals down to the bottom, e.g., by dropping the capillary repeatedly through a small funnel with a long pipe (a long glass tube alone also works).

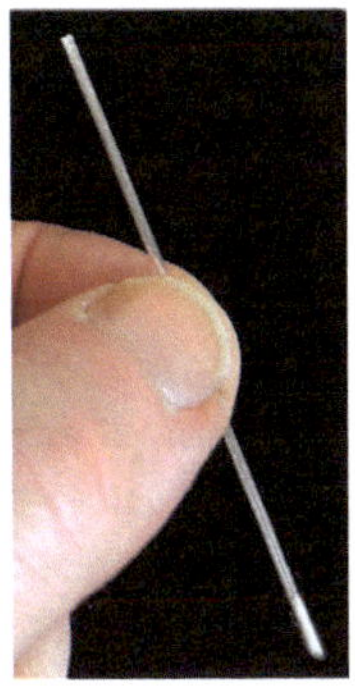

Figure 5.8c The crystals should now be packed at the closed bottom end of the capillary.

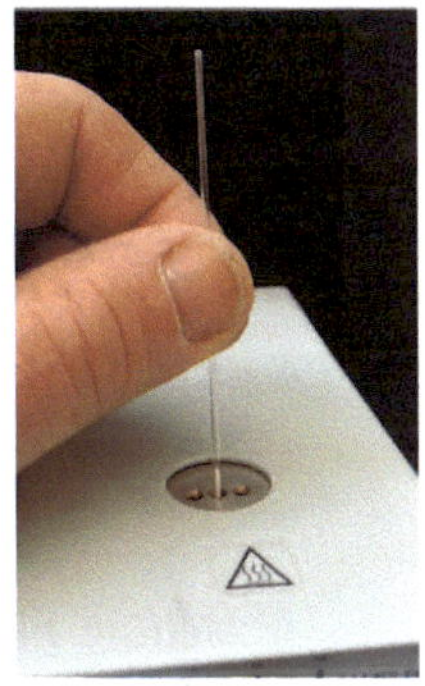

Figure 5.8d Insert the capillary into one of the sample holes of the apparatus. Make sure that the melting point apparatus has been recently calibrated (see Section 5.5.3).

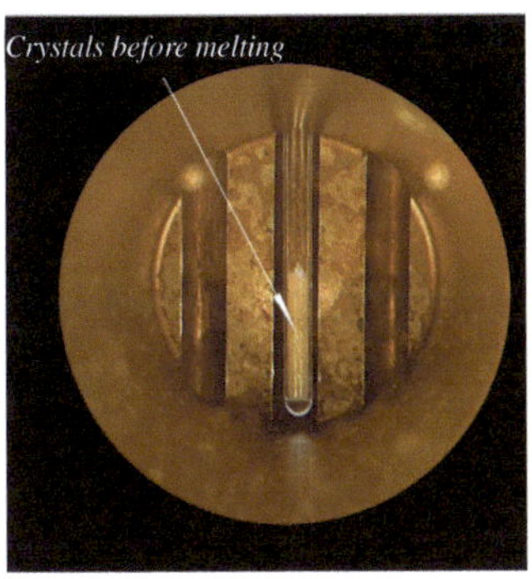

Figure 5.8e Looking through the apparatus magnifier, you should be able to see the capillary containing your crystals. If you know the approximate melting point of your crystals, program the apparatus to increase the temperature rapidly up to about 20 °C below that point, and then "idle" there.

Figure 5.8f Reprogram the apparatus to slowly increase the temperature (1 °C/min or less) while constantly observing the crystals. *Note the temperature at which the first signs of melting appear* (i.e., "sintering")!

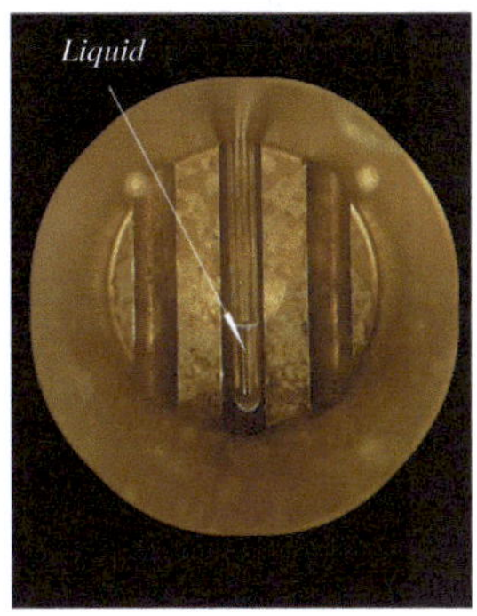

Figure 5.8g Eventually you will reach the temperature at which all solid is molten. *Note this temperature*! The "sintering" and "molten" temperatures are the "melting range." For pure crystals, this range is usually 1 °C or less. Impurities widen the range and shift it toward lower temperatures.

5.5.3 Tips and Tricks

1. If you do not know anything about the melting point of your crystals, you should first do a quick preliminary melting point determination by programming the apparatus to increase the temperature at an intermediate rate (4–5 °C/min) and note the approximate temperature at which your crystals melt. Then you can fill a new capillary and make a more careful determination.
2. *Getting the crystals to the bottom of the capillary (see Figure 5.8c), additional trick*: often the crystals remain packed in the opening of the capillary, no matter how much you tap and knock to get them down to the bottom. Vibrating the top part of the capillary with a file or the ribbed part of a tweezer can solve the problem.
3. *Melting point capillaries*: commercially available capillaries are usually 100 mm long with an inner diameter of 0.8–1 mm. One end should be open and the other sealed. If you only have capillaries with both ends open, you can easily seal one end by heating with a Bunsen burner until the opening closes.
4. *Melting point apparatus calibration*: several errors are possible when you determine a melting point: the capillary with the crystals may not experience the same temperature as the thermometer of the apparatus, the thermometer itself may show an incorrect value, etc. Therefore, the observed melting point/melting range of your crystals needs to be corrected. There is a quick and dirty way of doing this: in addition to determining the melting point of your crystals, you can determine the melting point of a “reference” compound with a known melting point in a similar melting range as your crystals. Compare the reference compound’s observed melting point to its known melting point, and use the difference between them to correct the observed melting point of your crystals.

A more accurate (and laborious!) way to correct melting points is to calibrate your melting point apparatus with an entire set of known reference compounds. After having determined the observed melting points for all of them, you can construct a calibration diagram with the observed and known (“literature”) melting points on the y- and x-axes, respectively. The diagram can then be used to correct all observed melting points determined with your apparatus. You simply enter the observed values for your crystals on the y-axis and read the corrected values on the x-axis (see Figure 5.9).

For the melting point determinations described above, you will need several milligrams of material. However, with a melting point microscope (Figure 5.10), you can determine melting points using only a few micrograms of material. The procedure is similar to the one given above, but you place a very tiny amount (>1 mg) of the crystals between two microscope glass slides rather than in a capillary, and you observe the melting process through a microscope rather than through a magnifying glass. Calibration is as important here as with a capillary-type apparatus.

Reference	litt. valu	obs. value
Phenyl salicylate	42	40
Benzophenone	48	45
Vanillin	82	79
Acetanilide	115	111
Phenacetin	134	125
Sulfanilamide	164	150
Caffeine	236	210

(a)

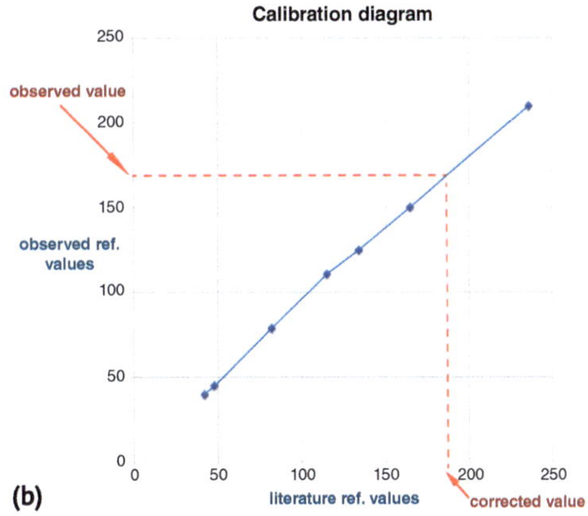

(b)

Figure 5.9 (a) Example of a calibration table and (b) diagram for a specific melting point apparatus.

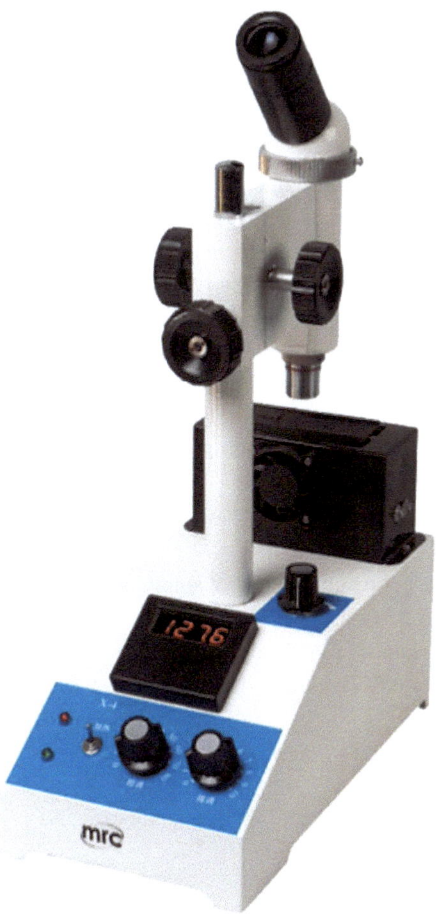

Figure 5.10 A melting point microscope. Courtesy of MRC Ltd.

References

1 Bienz, S., et al., *Spectroscopic Methods in Organic Chemistry*, 3rd Ed. 2021, Stuttgart: Georg Thieme Verlag KG. 512 pages.

2 Harwood, L.M. and T.D.W. Claridge, *Introduction to Organic Spectroscopy*. 1996, Oxford, UK: Oxford University Press. 96 pages.

3 Hesse, M., H. Meier, and B. Zeeh, *Spectroscopic Methods in Organic Chemistry*, 2nd Ed. 2008, Stuttgart: Georg Thieme Verlag KG. 419 pages.

4 Lambert, J.B., et al., *Organic Structural Spectroscopy*, 2nd International Ed. 2014, Essex, England: Pearson Education Limited. 464 pages.

5 Yadav, L.D.S., *Organic Spectroscopy*. 2004, Dordrecht: Springer. 324 pages.

6 Silverstein, R.M., et al., *Spectrometric Identification of Organic Compounds*, 8th Ed. 2014, Hoboken, NJ: John Wiley & Sons. 464 pages.

7 Günther, H., *NMR Spectroscopy: Basic Principles, Concepts, and Applications in Chemistry*, 3rd Ed. 2013, Weinheim: Wiley-VCH. 734 pages.

8 Friebolin, H., *Basic One- and Two-Dimensional NMR Spectroscopy*, 5th, Completely Revised and Updated Ed. 2010, Weinheim: Wiley-VCH. 442 pages.

9 Zerbe, O. and S. Jurt, *Applied NMR Spectroscopy for Chemists and Life Scientists*. 2014, Weinheim: Wiley-VCH. 548 pages.

10 Claridge, T.D.W., *High-Resolution NMR Techniques in Organic Chemistry*, 3rd Ed. 2016, Amsterdam: Elsevier. 552 pages.

11 Keeler, J., *Understanding NMR Spectroscopy*, 2nd Ed. 2010, Chichester, UK: John Wiley & Sons. 526 pages.

12 Lambert, J.B., E.P. Mazzola, and C.D. Ridge, *Nuclear Magnetic Resonance Spectroscopy: An Introduction to Principles, Applications, and Experimental Methods*, 2nd Ed. 2019, Hoboken, NJ: John Wiley & Sons. 480 pages.

13 de Hoffmann, E. and V. Stroobant, *Mass Spectrometry: Principles and Applications*, 3rd Ed. 2007, Weinheim: Wiley-VCH. 512 pages.

14 Watson, J.T. and O.D. Sparkman, *Introduction to Mass Spectrometry: Instrumentation, Applications, and Strategies for Data Interpretation*, 4th Ed. 2007, Hoboken, NJ: John Wiley & Sons. 848 pages.

15 McMaster, M.C., *LC-MS: A Practical User's Guide*. 2005, Hoboken, NJ: Wiley-Interscience. 157 pages.

16 Kind, T. and O. Fiehn, Advances in structure elucidation of small molecules using mass spectrometry. Bioanal Rev, 2010. 2: pp. 23–60.

17 Thomas, M.J.K. and D.J. Ando, *Ultraviolet and Visible Spectroscopy: Analytical Chemistry by Open Learning*, 2nd Ed. 1996, Chichester, UK: John Wiley & Sons. 300 pages.

18 Stuart, B.H., *Infrared Spectroscopy: Fundamentals and Applications*. 2004, Hoboken, NJ: John Wiley & Sons. 248 pages.

19 McMaster, M.C., *HPLC: A Practical User's Guide*, 2nd Ed. 2007, Hoboken, NJ: Wiley-Interscience. 256 pages.

6
Documentation

6.1 General Considerations

Careful and continuous documentation of experimental activities is essential in all sciences. In organic chemistry, it means that you should describe every laboratory experiment with a level of detail that is high enough for it to be safely reproduced. It also means that all the data that you have recorded in connection with the experiment (e.g., monitoring or characterization data) should be available in such a way that it can be easily linked to the respective experiment.

Before the "computer age," almost all lab documentation was paper-based. Lab notebooks (Figure 6.1) were used for the continuous (usually chronological) documentation of experiments, and there were binders or folders where the associated data (e.g., MS or NMR spectra) could be found. The connection of the data associated with the notebook experiment was by batch numbers, book pages, or dates.

Computers (Figure 6.1b) are excellent tools for data storage and organization, and digital documentation has therefore, from 1980s onward, gradually replaced paper-based documentation in chemistry labs. The possibilities to search, overview,

(a) (b)

Figure 6.1 (a, b) Devices for laboratory documentation.

reorganize, store, duplicate, and transfer data are much better with digital documentation. In the following sections, the documentation of an "example experiment" will be shown in three different formats, one paper-based and two different digital formats.

6.2 Lab Notes Handwritten on Paper

"Paper and pen" were the usual tools for documentation of laboratory experiments until very recently, and such documentation is still common. Over the years, organic chemists have agreed on certain principles that should be followed:

- Preferably, a bound lab notebook should be used.
- The handwriting should be readable and produced with a pen, not a pencil.
- The notes should be organized chronologically, on numbered and dated pages.
- If the notes describe a preparation, the product should be given a unique batch number according to some consistent system (e.g., book page or date, with extension numbers in case there are several products).
- Data associated with the notes (e.g., monitoring and characterization data) should be either included or collected separately and then organized so that the connection to the lab notebook is clear (e.g., by batch number or book page).

The notebook text in Figure 6.2 (describing a short organic chemistry preparation) illustrates these principles. In addition to the items 1–10, many chemists include a separate table of all reagents and their amounts, much like in a cookbook recipe. The ^{1}NMR mentioned in the text was stored in paper form as "111-069-1" in a binder together with other spectra in numerical order. The TLC accompanying the text was a pasted-in photocopy of the actual plate.

6.3 Lab Notes on Electronic Media Created with Word-processing Software

Beginning in the 1970s, a gradual transition from paper to electronic media as carriers of information took place. Nowadays, the majority of chemistry labs create and store their documentation on computers, with regular backups being done either on local storage media or with the aid of cloud services. In the beginning of the computer era, the most common way to produce electronic lab notes was to use existing word-processing software and build the documentation around text files created with this software. Very early on, it had also become possible to paste pictures, graphs, and chemical structures into the documents, enabling almost all lab documentation to be squeezed into one single word processor file. An example of what such a document could look like is given in Figure 6.3. It describes the same experiment as in Figure 6.2, but now in a more readable form. The inserted TLC graphic, now a digital photo, is also more detailed. The chemical formulas were

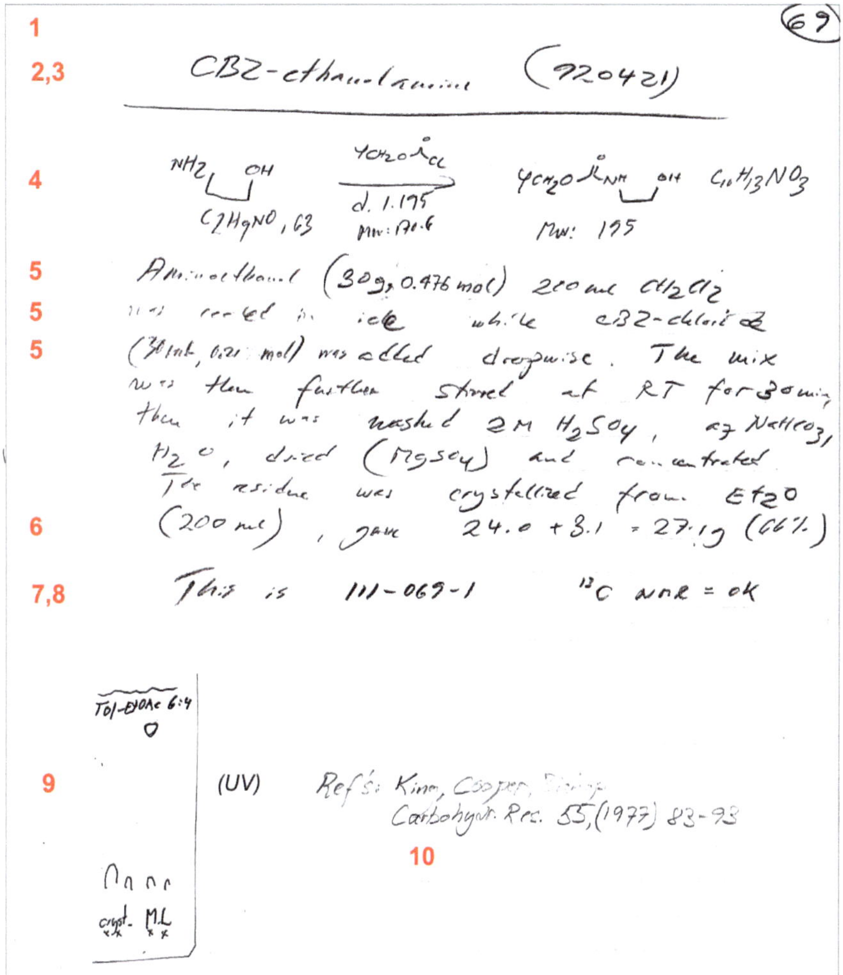

Figure 6.2 Photocopy of page 69 in notebook # 111. The handwriting is not the best, but (hopefully) readable. The red numbers indicate the data that should be included: 1: page number, 2: title, 3: date, 4: reaction formula, 5: descriptive text with reactants (weights and moles) and solvent (volume), 6: yield of product (actual weight and mol.%), 7: product batch number, 8: characterization (here NMR), 9: monitoring (here TLC, with eluant and visualization specified), and 10: relevant literature reference(s).

created with a chemical drawing program and inserted into the word processor file. The NMR spectrum was stored as a separate file and named so that its association with the text was clear.

Word processors can also provide templates that facilitate uniformity of lab notes, which is very practical in a company or research group. If a dominating word processor (e.g., MS Word, Keynote, and OpenOffice) is used, the compatibility with future computer environments and operative systems is secured (an important aspect!).

CBZ-Ethanolamine (920421)

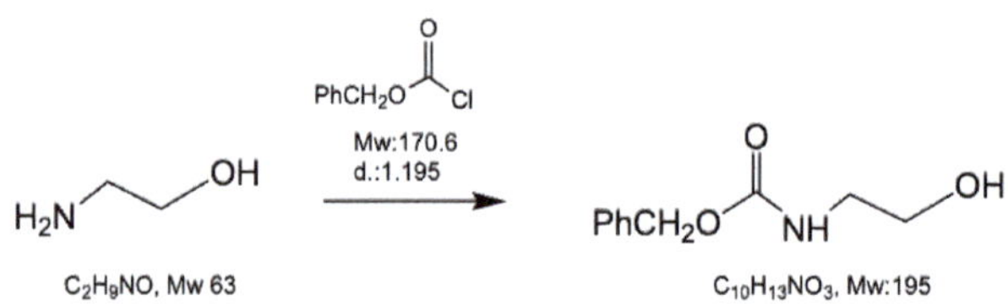

Aminoethanol (30 g, 0.476 mol) in CH_2Cl_2 (200 mL) was cooled in ice while CBZ-chloride (30 mL, 0.21 mol) was added dropwise. The mix was further stirred at RT for 30 min, then it was washed with 2M H_2SO_4, aq. $NaHCO_3$, H_2O, dried ($MgSO_4$) and concentrated. The residue was crystallized from Et_2O (200 mL), this gave 24.0 + 3.1 = 27.1 g (66 %, theor. 40.95 g), this is 111-069-1, ^{13}C-NMR was OK.

(UV)

Ref's: R. R. King, F. P. Cooper and C. T. Bishop, Carbohydrate Research 55 (1977) 83-93.

Figure 6.3 The same experiment that was "hand-documented" in Figure 6.2, now documented in electronic form: a single word processor file with inserted reaction scheme and TLC photo file.

6.4 Lab Notes on Electronic Media Created with Specialized ELN (Electronic Lab Notes) Software

Although digital lab documentation can be created and retrieved with the aid of word processors, there is a growing need, especially in larger organizations, for further standardization of the way that documentation is created, organized, and retrieved. Therefore, a plethora of specialized applications have emerged that integrate experimental text, structural data, spectral data, etc. in one database file. Presently, more than 50 different ELN applications are found on the market, making a review of even a few of these too ambitious, at least here. Common to all ELN applications is that the data entered by the user is stored in a central database, either site- or cloud-based. Descriptive text is entered much like in an ordinary word processor, with the added option that extra data connected to the experiment (spectra, chromatograms, etc.) can be inserted in a simple way. Many ELNs also calculate or

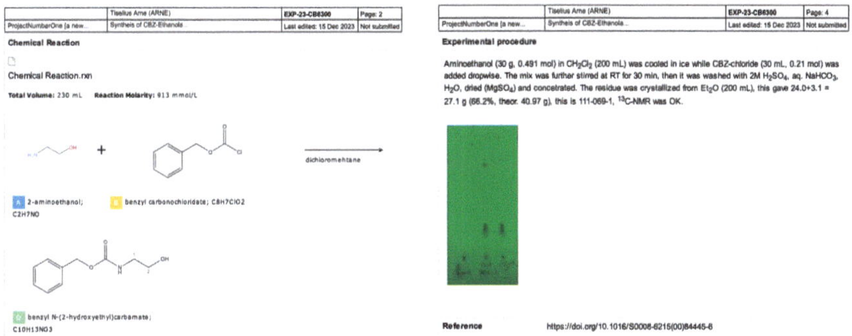

Figure 6.4 Two screenshots from an ELN entry of the same example experiment as in Figures 6.2 and 6.3. In addition to the screens shown, other screens with data for the reactants and product(s) (molecular weights, purity, sum formulas, etc.) can be displayed, as well as screens showing reference paper text or spectral data in pdf format. Some ELNs provide the option to append spectra in their original instrument "raw" file format.

retrieve data like molecular weight or safety data for all relevant chemicals, which is very time-saving. Figure 6.4 shows how the previous "example experiment" could look like in an ELN environment.

From an employer/administrator perspective, the advantage of ELNs is that all user data are stored in a common database which can be quickly searched, and experimental data quickly retrieved. The legal aspects of lab documentation are also well taken care of since both distribution and access to ELN data are controlled by the employer/administrator. Possible drawbacks are the associated costs for software and database maintenance.

From the chemist/user perspective, an ELN application can, once learned and mastered, significantly reduce the time needed for fetching chemical data, calculating correct reactant amounts, and draw structures. Possible drawbacks are high costs, increased dependance on site or cloud servers, and technical difficulties when exporting notebook data to another ELN application or to other digital or physical formats.

7

Safety in the Laboratory

7.1 Introduction

Laboratory work is not without certain risks, but, as with all other risky undertakings, knowledge and suitable protective measures make a big difference. In the following, we will discuss laboratory safety rules, personal protective equipment, fixed laboratory safety equipment, and some facts about dangerous chemicals and their handling. However, laws and regulations for lab work vary between countries, and each lab and employer have their own set of safety rules and regulations. It is the chemist's duty to be familiar with rules and regulations that apply to the specific workplace and follow them. This chapter discusses only general safety issues and rules.

Very often, your local set of safety rules are written with so much detail and contain so much information that it could be difficult to sort out the most important issues. This is unfortunate and counterproductive, since there are a few "most important" general rules that apply to all organic chemistry lab works. These rules are summarized below as the "10 commandments" and, just like in the Bible, the most important commandment is the first one.

The 10 commandments of organic chemistry lab work:

1. Always wear safety goggles!
2. Always wear other recommended personal safety items (lab coat, proper shoes, and proper hair arrangement).
3. Always be familiar with the location and use of fixed safety arrangements (e.g., firefighting equipment, emergency shower, and eyewash equipment) and the emergency exits.
4. Never start any preparative work before you have informed yourself about the risks associated with every chemical and apparatus that you are going to use.
5. Always run reactions (and most workup procedures) in a well-ventilated fume hood.
6. Never eat or drink in the lab.

7. Never perform preparative work alone.
8. Always put chemicals back in their place as soon as you have used them.
9. Always label flasks and vials, even temporary ones (use a felt-tip pen or similar for the latter).
10. Take proper notes during your experiments.

Below are some further comments about the 10 commandments and other safety issues. Regarding commandment number 10, Chapter 6 describes how to take proper lab notes.

7.2 Personal Protection

7.2.1 Safety Goggles

This is the single most important personal protective item in the organic chemistry laboratory! Its importance cannot be overstressed. Always wear safety goggles in the lab! Acid splashes or shattered pieces of glass can damage your lab coat and clothes or even give you small scars, but if they hit your eyes you may lose your vision and be seriously handicapped for the rest of your life!

There are quite a few models of plastic protective goggles available from chemical vendors. The ones shown in Figure 7.1a–c are suitable for students in a course lab. They are lightweight, durable, and provide protection in both the forward and sideways directions. The larger versions can accommodate ordinary goggles on the inside (Figure 7.1c).

For organic chemists that spend many hours every day in the lab for longer periods (PhD students, postdocs, and industry professionals), the recommendation is to use personalized protective goggles from an optician. They will provide a better fit, both physically and optically, and there is a wider choice of models and materials. The employer often covers the cost for such goggles, since it is a good and reasonable investment in staff safety.

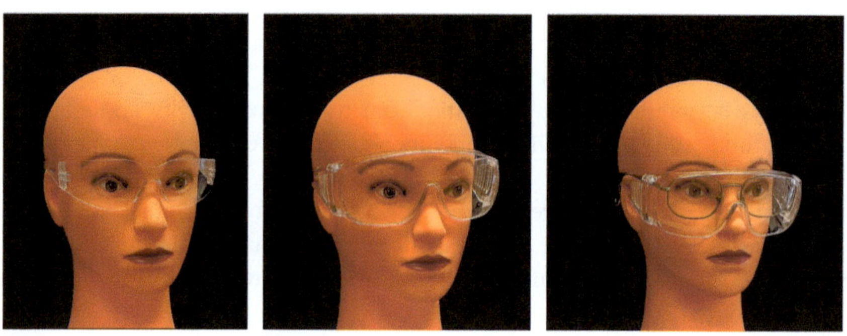

(a) (b) (c)

Figure 7.1 Different plastic goggle models suitable for use in, for example, course labs.

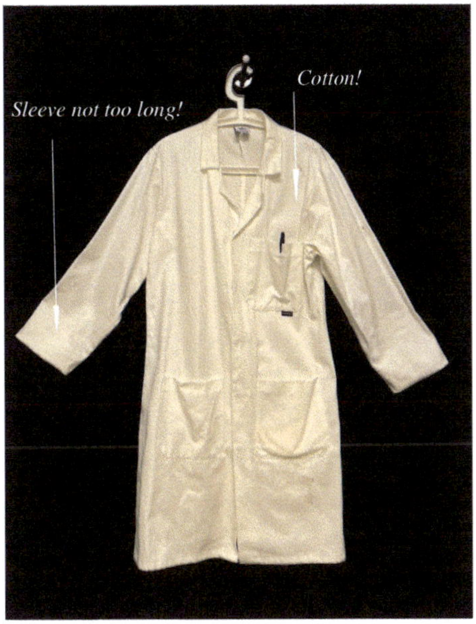

Figure 7.2 Typical cotton lab coat. This one has "snap" buttons for "fast off" (important if it catches fire!).

7.2.2 Lab Coat

The second most important personal protection item is the lab coat (Figure 7.2). It provides an extra layer of protection from spills and even small fires. Note that cotton is the recommended lab coat material. Coats made of synthetic materials (e.g., nylon) can melt when burning and give very serious wounds and should therefore be avoided. It is also important that the lab coat fits you well and does not have too long and wide sleeves (which can accidentally tip smaller flasks over, causing spills and accidents). As with protective goggles, lab coats are often provided by the employer.

7.2.3 Gloves

When working with chemicals in the lab, gloves should be worn when deemed necessary, and they must be made of a material that really protects against the chemicals handled, see the (unfortunately too detailed) chart in Figure 7.3. A typical example is latex gloves when handling bromine; in this case gloves are always recommended. When handling other, less hazardous chemicals, the question of whether one should use gloves or not is more difficult to answer. As always, scale matters. In a pilot-plant lab that frequently handles kilograms of chemicals, it may be necessary to use gloves all the time in the lab. In a small-scale research lab, however, routine use of gloves is unusual and even somewhat controversial. Presence of spills on a gloved hand is more difficult to sense than on a bare hand, and this

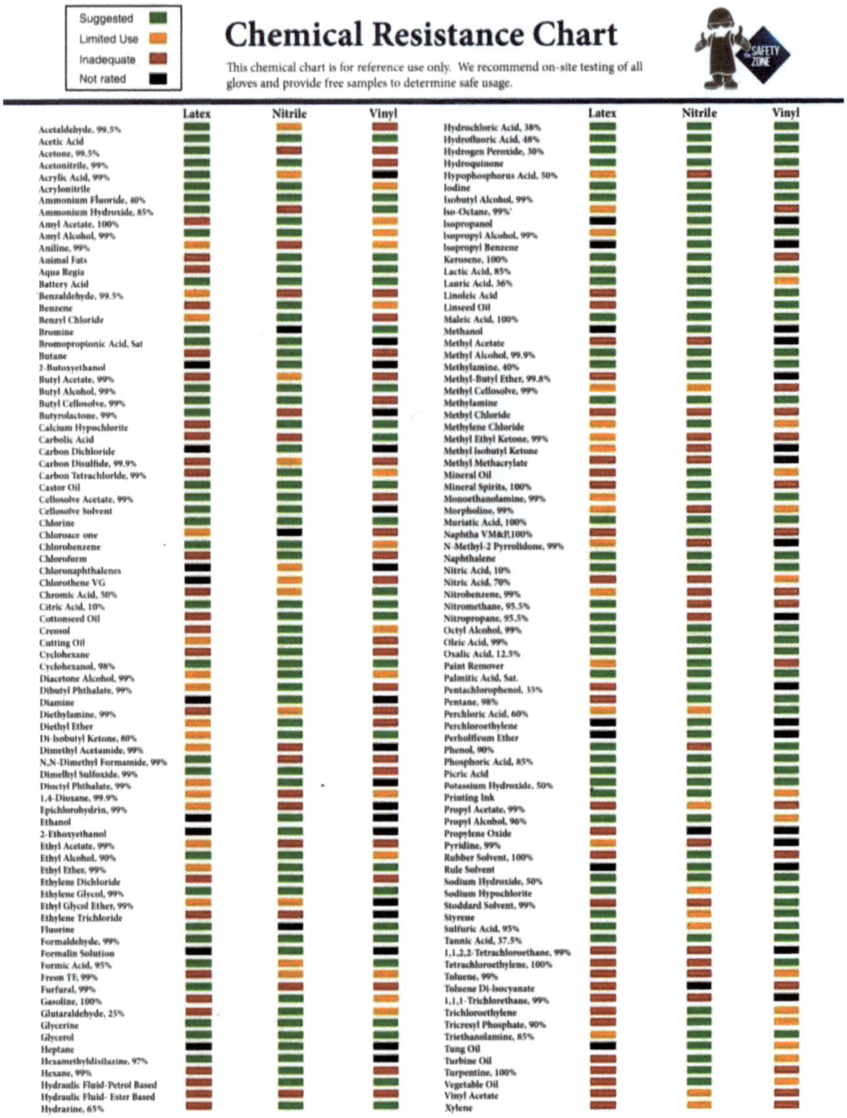

Figure 7.3 Chemical resistance of glove material.

could lead to unnecessary contamination of laboratory equipment, benches, or door handles. Furthermore, some of the "tactile" sensitivity is lost when you use gloves, no matter how thin they are. This, together with the increased "slipperiness" of wet gloves, could lead to more fumbling during benchwork.

If you are working without gloves in the lab, be very observant of spills on your hands! Rinse immediately if you see or sense a spill. Use plenty of water first, then the appropriate solution or solvent (ethanol for water-immiscible liquids is common). With acid spills, it is a good practice to first rinse with water, then make a short rinse with 1 M aqueous sodium bicarbonate, and finally rinse with water. Bottles with bicarbonate solution are found on the shelf in almost every lab. Similarly, alkali spills are rinsed first with water, then with dilute acetic acid, followed by again lots of water.

7.2.4 Shoes

Many labs do not allow "open-toe shoes," i.e., sandals or flip-flops. The reason is that spills on your feet often pass undetected and can cause burns later. However, lab work involves a lot of walking and standing for many hours, and this puts a lot of strain on your feet and legs. Sandals of the "orthopedic" type can then be a better solution than, for example, sneakers and other shoes that have bad support and/or bad breathing properties. As always, the type of lab work, the chemicals involved, and the scale of the activities matter. Follow the local rules!

7.3 Fixed Laboratory Safety Equipment

7.3.1 Fume Hood

A well-ventilated fume hood is the single most important fixed safety item! That is where you should carry out all your reactions and most of your workup procedures as well. Some important features are highlighted in Figure 7.4. Note that the fume hood in the picture shares its ventilation duct with the surrounding cabinets so that the entire surrounding storage space becomes constantly ventilated.

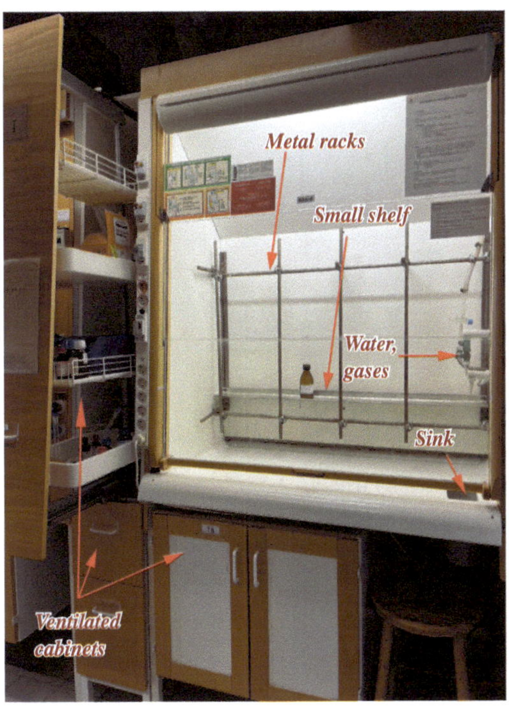

Figure 7.4 A modern fume hood surrounded by ventilated drawers and cabinets.

7.3.2 Firefighting Equipment

Every organic chemistry lab should have firefighting equipment easily available. The most important item is a fire extinguisher of the dry ice type (Figure 7.5). It is very important that everyone in the lab knows where to find it and how to use it. The filling status and function should be periodically checked.

7.3.3 Gas Cylinders

Most labs use high-pressure gas cylinders as source of, for example, argon, nitrogen, or hydrogen. The best solution is to store the gas cylinders in specialized steel cabinets, with steel tubes leading to fume hood outlets. However, in labs that use only nitrogen or argon, you often see a gas cylinder placed next to the fume hood, with a rubber tubing leading into the hood. This arrangement is dangerous! The cylinder (tall and often very heavy) can fall, causing personal injury or tearing items out of the

Figure 7.5 Dry ice fire extinguisher.

hood. In the worst case, the gas regulator vent can break off in the fall and the cylinder can become a gas-propelled rocket! Therefore, gas cylinders in the lab *must* always be secured (with a chain or similar) to a wall or other lab construction element.

7.3.4 First-aid Kit

Every lab should have a simple first-aid kit permanently placed somewhere in a clearly visible place, e.g., mounted on a wall (Figure 7.6). Cuts from broken glass,

Figure 7.6 Wall-mounted first-aid kit.

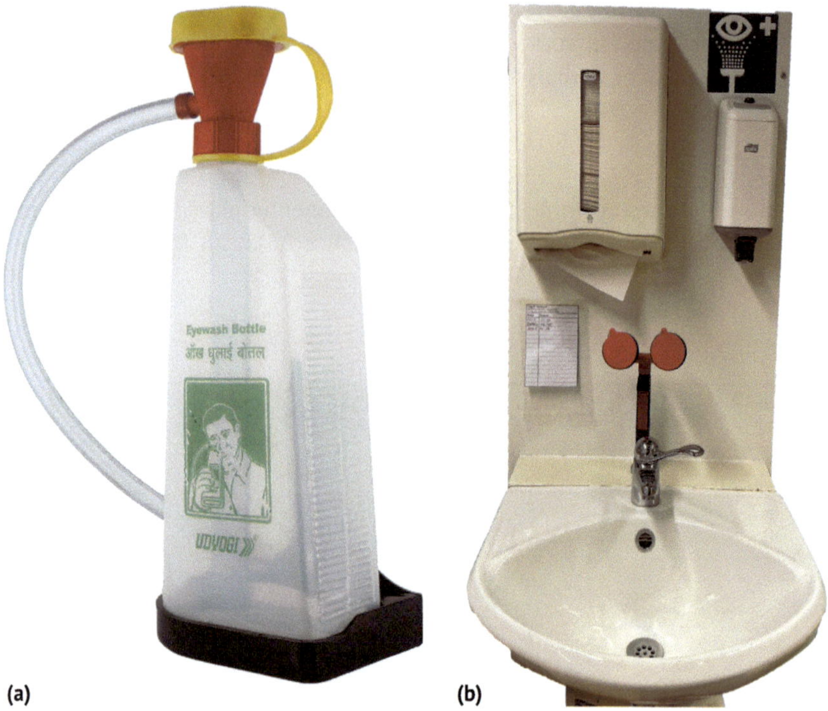

(a) (b)

Figure 7.7 (a) Eye-wash bottle. (b) Eye-wash station.

especially on hands and arms, are the most common injuries in the lab. A kit should, at a minimum, include band-aids of different forms and sizes, and a disinfectant solution.

7.3.5 Eye-wash Equipment

Eyes protected by well-fitting safety goggles are still not 100% protected against splashes from all directions, so if you sense or suspect some eye contamination, immediately rinse your eye with plenty of water, preferably with an eye-wash bottle (Figure 7.7a) or at an eye-wash station (Figure 7.7b). Every lab should have such equipment in a clearly visible place, and the function should be periodically checked.

7.4 Dangerous Properties of Reagents and Solvents

It cannot be enough stressed: it is every chemist's personal responsibility to know enough about the properties of the reagents, solvents, and equipment that he/she intends to use. To aid in this, older practical organic chemistry handbooks contained large sections describing hazards and safe handling of common solvents and chemicals, see, for example, Vogel's [1] or Leonard's [2] handbooks. There are also dedicated publications on chemicals and safety, see, for example, [3–5]. However, a listing of dangerous chemicals and solvents and their handling goes beyond the scope of this book. Such information can be retrieved from the references indicated above or from chemical vendors' websites. Most such sites nowadays present hazardous chemical information in the standardized, internationally agreed GHS format. GHS stands for "Globally Harmonized System of Classification and Labelling of Chemicals."

The GHS standard "defines and classifies the physical, health, and/or environmental hazards of a substance." Each "class" has a pictogram associated with it (Figure 7.8). The GHS also defines how the SDS (Safety Data Sheet) for each substance should be formatted and what type of information should be included. It is from the SDS (formerly MSDS) that you can get detailed information about a substance such as stability, toxicity, and flammability, among others, how to safely handle it, what to do if there is an accident (fire, spill, ingestion, etc.), how to store the substance, how to dispose of it, plus a lot of other useful information. If you are a beginner, the SDS (often 10+ pages!) may appear too detailed and fact-filled, but once you get used to the (always uniform) format you learn to "extract" the information that applies to your specific situation.

Figure 7.8 GHS pictograms.

The entire SDS of diethyl ether here serves as an example. The sections of the 12-page document are:

- Section 1: identification of the substance/mixture and of the company/undertaking
- Section 2: hazards identification (entire contents shown in Figure 7.9)
- Section 3: composition/information on ingredients
- Section 4: first-aid measures
- Section 5: firefighting measures
- Section 6: accidental release measures
- Section 7: handling and storage
- Section 8: exposure controls/personal protection
- Section 9: physical and chemical properties
- Section 10: stability and reactivity
- Section 11: toxicological information
- Section 12: ecological information
- Section 13: disposal considerations
- Section 14: transport information
- Section 15: regulatory information
- Section 16: other information

From this list, we have expanded Section 2, "hazards identification," in Figure 7.9.

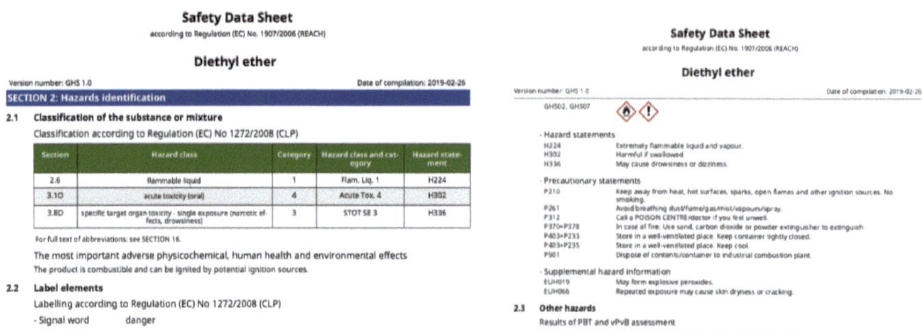

Safety Data Sheet

according to Regulation (EC) No. 1907/2006 (REACH)

Diethyl ether

Version number: GHS 1.0 Date of compilation: 2019-02-26

SECTION 2: Hazards identification

2.1 Classification of the substance or mixture

Classification according to Regulation (EC) No 1272/2008 (CLP)

Section	Hazard class	Category	Hazard class and category	Hazard statement
2.6	flammable liquid	1	Flam. Liq. 1	H224
3.10	acute toxicity (oral)	4	Acute Tox. 4	H302
3.8D	specific target organ toxicity - single exposure (narcotic effects, drowsiness)	3	STOT SE 3	H336

For full text of abbreviations: see SECTION 16.

The most important adverse physicochemical, human health and environmental effects

The product is combustible and can be ignited by potential ignition sources.

2.2 Label elements

Labelling according to Regulation (EC) No 1272/2008 (CLP)

- Signal word danger

- Pictograms

GHS02, GHS07

- Hazard statements

H224 Extremely flammable liquid and vapour.
H302 Harmful if swallowed.
H336 May cause drowsiness or dizziness.

- Precautionary statements

P210 Keep away from heat, hot surfaces, sparks, open flames and other ignition sources. No smoking.
P261 Avoid breathing dust/fume/gas/mist/vapours/spray.
P312 Call a POISON CENTRE/doctor if you feel unwell.
P370+P378 In case of fire: Use sand, carbon dioxide or powder extinguisher to extinguish.
P403+P233 Store in a well-ventilated place. Keep container tightly closed.
P403+P235 Store in a well-ventilated place. Keep cool.
P501 Dispose of contents/container to industrial combustion plant.

- Supplemental hazard information

EUH019 May form explosive peroxides.
EUH066 Repeated exposure may cause skin dryness or cracking.

2.3 Other hazards

Results of PBT and vPvB assessment

According to the results of its assessment, this substance is not a PBT or a vPvB.

Figure 7.9 The "Section 2" part ("hazards identification") for diethyl ether in its entire form. Most of the important safety information is summarized here.

Finally, and *very* importantly, a word about scale when working with dangerous chemicals and solvents: the risks increase dramatically with the scale of operation! A semi-microscale (0.1–1 g) pyrophoric reagent experiment that goes wrong may cause a small fire in the fume hood which can easily be extinguished before it spreads. If the same experiment on a 200 g scale goes wrong, the fireball may knock you over, burst out of the fume hood, and ignite the entire lab before you even start thinking about the fire extinguisher!

It is the same story with toxic reagents: you may successfully avoid getting your bare hands contaminated if you handle fractions of a gram with a spatula. However, if you have to “shovel” several hundred grams with a scoop held in your bare hand, you will sooner or later spill some of the toxic stuff on your hand and wish that you had used a proper glove.

The take-home message: the larger the scale, the larger the risks and need for careful precautions! Also, to avoid surprises when scaling up reactions, the rule of thumb is to proceed stepwise, with no more than a 5× increase of scale each time.

References

1 Vogel, A.I., et al., *Vogel’s Textbook of Practical Organic chemistry*, 5th Ed. 1989, Essex, England: Longman Scientific & Technical and John Wiley & Sons. 1514 pages.

2 Leonard, J., B. Lygo, and G. Procter, *Advanced Practical Organic Chemistry*, 3rd Ed. 2013, Boca Raton, FL: CRC Press. 356 pages.

3 Royal Society of Chemistry, *Safe Practices in Chemical Laboratories*. 1989, London, UK: The Royal Society of Chemistry. 59 pages.

4 Lewis, R.J., Sr., *Sax’s Dangerous Properties of Industrial Materials*, 11th Ed. 2004, New York: John Wiley & Sons. 4860 pages.

5 Furr, A.K., *CRC Handbook of Laboratory Safety*, 5th Ed. 2000, New York: CRC Press. 802 pages.

8

Other Useful Information

8.1 General Considerations

As noted already in Chapter 1, synthetic organic chemists can accomplish quite remarkable things using relatively simple equipment. Only the analytical instruments used for product characterization are sophisticated and expensive. The first basic operations during a synthetic step, reaction setup/running and monitoring (Chapters 2 and 3), and reaction workup (Chapter 4) require only simple laboratory glassware and some devices that can weigh, stir, heat, cool, clean, store, or produce reduced pressure ("vacuum"). All these devices are comparatively simple and affordable, and together they make up the "basic tools" that are found in almost every organic chemistry laboratory. Knowledge about these, their function, handling, and maintenance is essential for smooth and efficient organic synthesis work. The purpose of this chapter is to provide you with some of that knowledge. In contrast to the earliest chapters, where step-by-step instructions were frequent, this chapter is mostly "static" and descriptive.

8.2 Vacuum Sources

8.2.1 General

Organic chemistry lab work often requires a vacuum source, e.g., for suction filtration, rotavaping, or vacuum distillation. Several different types of pumping devices are used, depending on the type of vacuum needed and local conditions. Below are some examples.

8.2.2 Water Aspirator Pumps (10–20 mbar)

For a "moderate" vacuum (filtrations, evacuations, and rotavapor evaporations), a water aspirator pump (Figure 8.1) is a simple and well-known solution. In this device (made of glass, metal, or plastic), compressed water from the tap is

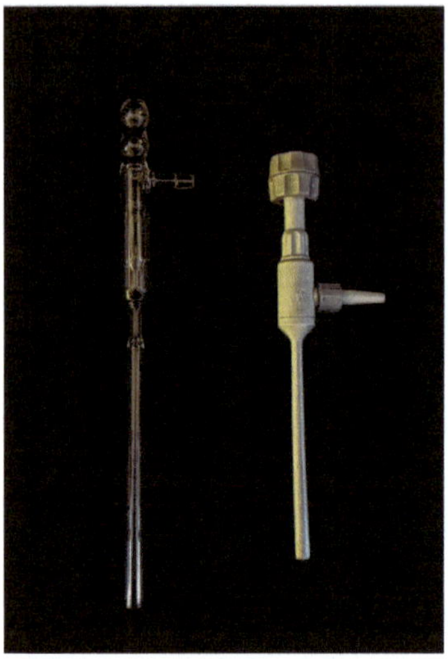

Figure 8.1 Water aspirators.

forced through a narrow nozzle, creating an air suction effect. A “vacuum” of 10–20 mbar can be obtained, depending on the tap water temperature and pressure. The water aspirator pump is by far the simplest and cheapest vacuum source! However, when connected to an ordinary tap water supply, it is associated with a water consumption cost. In some regions of the world this cost is considerable, and for this and other reasons (e.g., unfavorable water temperature or pressure, prohibitive legislation), simple water aspirators may not always be the best for the job.

Regardless of local conditions, the exhausts from water aspirators, if used in combination with rotavapors for long periods, could carry considerable amounts of volatile organic solvents into the municipal sewage system. For this reason, water aspirators are today not recommended for use with rotavapors. However, for processes such as occasional filtrations, they remain the simplest and cheapest solution.

A final word of caution regarding water aspirators, whatever purpose you use them for: they can “back-suck” water into your system if the tap water pressure suddenly drops or if you close the water tap too early. Always use a “safety bottle” between the aspirator and your equipment (Figure 8.2), and perhaps also an “anti-suck-back” valve as an extra precaution.

Aspirators can also be used together with a water container and an electric pump that recirculates and pressurizes the water. This circumvents the water consumption and solvent contamination problem [1]. Such “closed” devices are much more expensive than simple water aspirators, but they are still common in labs where tap water is unsuitable.

Figure 8.2 Vacuum spillover "safety bottle."

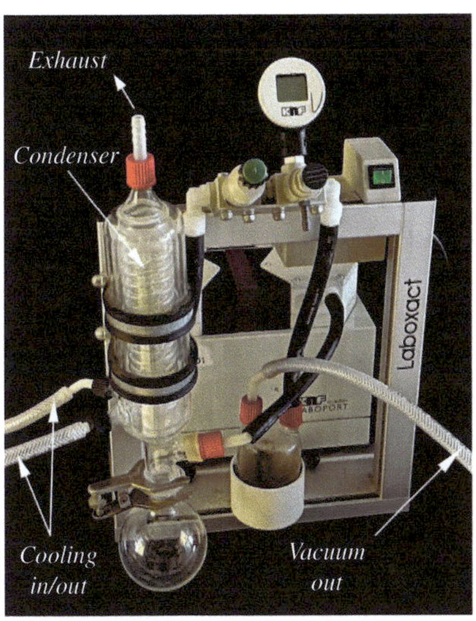

Figure 8.3 A diaphragm vacuum pump. The condenser and flask collect emitted solvent vapors.

8.2.3 Diaphragm Pumps ("Membrane Pumps," 10–20 mbar)

Next step up (pricewise) from a water aspirator is a two-stage diaphragm pump capable of providing a "vacuum" suitable for most rotavapor evaporations and many vacuum distillations (Figure 8.3). In this device, an electric motor moves two (or more) diaphragms back and forth, and together with one-way valves this creates an air pumping effect. A scrubber/condenser for the exhausts is also usually included. The fact that these pumps do not require vacuum oil (like rotary vane oil pumps) and that the housing, diaphragm, and valves can be made of corrosion-resistant materials makes them a very attractive choice for today's organic chemistry lab. They are also much easier to service than rotary vane oil pumps. The price level is similar to that of rotary vane pumps, but once purchased a diaphragm pump will probably survive heavy lab use for a longer time, if properly serviced.

8.2.4 Rotary Vane Oil Pumps ("Rotary Pump," "Rotary Oil Pump," 10^{-2}–5 mbar)

This is still the most common type of vacuum pump in chemistry labs. On the inside it holds a "rotating vane," partly immersed in silicon oil, which creates a strong air suction effect, achieving a lower pressure ("better vacuum") than a water aspirator or diaphragm pump. Such pumps (see Figure 8.4) are used in, for example, demanding vacuum distillations, Schlenk manifolds, and freeze-dryers.

Figure 8.4 A small rotary vane vacuum pump.

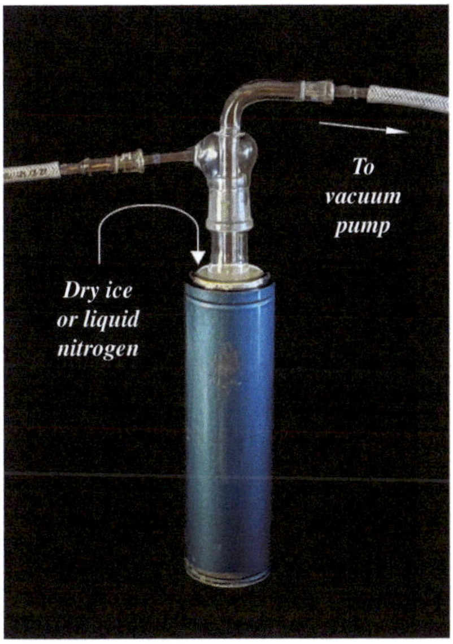

Figure 8.5 A laboratory cold-trap.

Compared to water aspirator and diaphragm pumps, rotary pumps are more sensitive to corrosive environments and therefore need more precautions and maintenance. During operation, a "cold-trap" (see Figure 8.5) should always be installed between the pump and the equipment served. Oil changes at regular intervals are also necessary, the time between changes depends on the type of use. Constantly operating freeze-dryer pumps need oil changes every second or third month.

8.2.5 Oil Diffusion Pumps (10^{-10}–10^{-2} mbar) and Turbomolecular Pumps (10^{-10}–10^{-1} mbar)

These pumps are capable of generating extremely low pressures if they are used in combination with "pre-evacuating" pumps. Examples of uses are in high-vacuum distillation setups (oil diffusion pump) and mass spectrometers (turbomolecular pumps). A more detailed description of these pumps is outside the scope of this book.

8.2.6 House Vacuum

Many laboratories have installed a "house vacuum" system where a large vacuum pump, with its associated pressure regulator and cold-trap, serves many outlets in the lab. The "vacuum" available can be anywhere from 50–100 mbar, suitable only for simple operations like filtration, to 5–10 mbar suitable for most evaporations and distillations. If used with proper caution and discipline, a house vacuum system (a sizeable investment) can save money in the long term for a medium- or large-sized laboratory. The disadvantages are dependance on one single device for an entire lab and the pressure fluctuations that can occur when many users suddenly open or close their connections.

8.2.7 Tips and Tricks

8.2.7.1 Pressure Unit Confusion

The unit "Torr" is often used; 1.00 Torr is equal to 1.33 mbar, which is equal to 1.33 hPa (hectopascal), which is equal to 0.0193 psi. Pa is the standard SI unit.

8.2.7.2 Cold-traps

As mentioned above, many pumps need a "cold-trap" installed upstream to protect their interior parts from corrosive or toxic vapors. In the lab, cold-traps usually consist of a glass structure immersed in a Dewar thermos filled with cooling medium (dry ice/ethanol or liquid nitrogen), an example is shown in Figure 8.5. Various electric low-temperature devices are also available, but they have a sizeable price tag. In freeze-dryers, compressor-driven cooling coils inside a large metal chamber efficiently "trap" the sample vapors before they reach the vacuum pump.

A word of caution regarding cold-traps immersed in liquid nitrogen (boiling point: −196 °C): if a steady flow of air passes through such traps, condensation of significant amounts of air oxygen (boiling point: −183 °C) could occur. If flammable organic solvents also mix in, there is a significant fire/explosion risk! Empty the trap often and do not restart the cooling until it is evacuated.

8.2.7.3 Vacuum Gauges

Whenever vacuum is used, it helps to keep track of the actual pressure, continuously or periodically. Stationary equipment like Schlenk lines or rotavapors should be connected to a vacuum gauge, and a portable "maintenance gauge" should also be available in the lab for occasional use. Most lab vacuum gauges are of the electronic Pirani type (10^{-3}–1 mbar range, Figure 8.6). For less demanding applications, a cheap mechanical gauge (20–1000 mbar range, Figure 8.7) can be sufficient. Mercury-based vacuum gauges were standard in 19th and early 20th century labs, but they have now been phased out because of the health risks associated with mercury.

8.2.7.4 Vacuum Tubing and Leaks

Regardless of the type, your equipment has to be connected to the vacuum source by some sort of tubing. Laboratory vacuum tubing is usually made of rubber or reinforced PVC, see Figure 8.8. The latter tubing is actually intended for compressed air, but it works well also for vacuum applications. It is cheap and readily available in, for example, automotive stores. It is also more durable than rubber tubing, which cracks/hardens rather quickly in a lab environment. However, reinforced PVC tubing can be quite stiff and tricky to handle. Prewarming with a heat gun or in a hot water bath helps!

Figure 8.6 Pirani vacuum gauge.

Figure 8.7 Mechanical vacuum gauge.

When your equipment has been connected to its vacuum source and your vacuum gauge does not display the expected pressure, you should suspect a leak. If you have not done it before, smear all joints and stopcocks with vacuum grease, use hose clamps (again, from automotive stores), and check again. Leak detection and -proofing are nearly unavoidable when working with vacuum equipment.

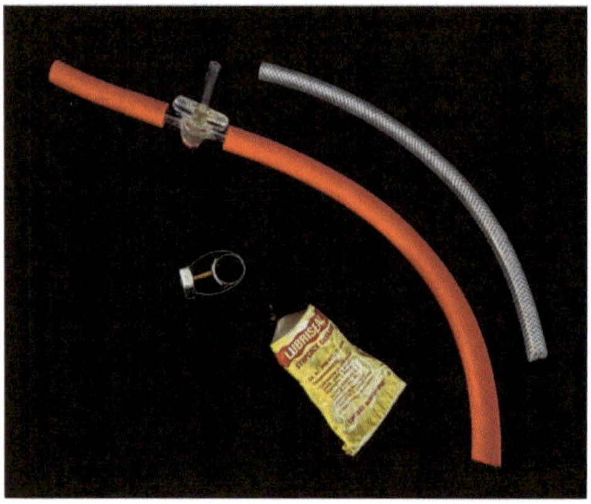

Figure 8.8 Vacuum tubing made of reinforced PVC (upper) and rubber (lower), hose clamp, and vacuum grease.

8.2.7.5 Manifolds and Permanent Vacuum Setups

Many organic chemists use both vacuum and inert gas very frequently. For convenience, they often set up a fume hood manifold arrangement that enables them to work with both together. A popular variant is the "Schlenk line" (Figure 8.9). It consists of, from left to right, a vacuum source (e.g., a rotary pump; Figure 8.4), a vacuum gauge (Figure 8.6), a cold-trap (Figure 8.5), a double manifold (Figure 8.10),

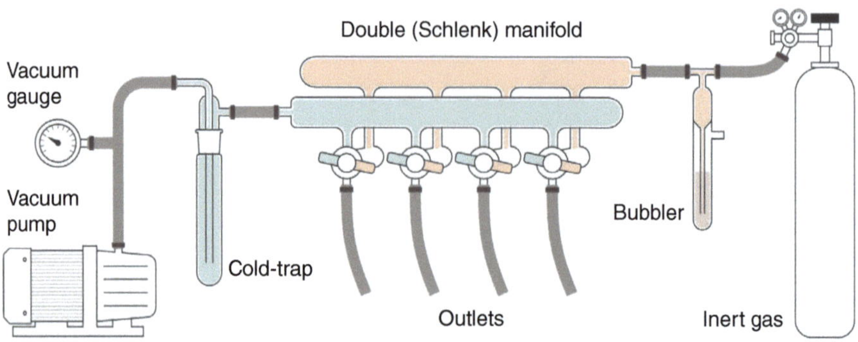

Figure 8.9 Schematic representation of a Schlenk line setup. Tubing (here colored black) is made of rubber or PVC. If necessary, an additional purification device can be fitted between the inert gas source and the bubbler. Courtesy of Margret Holtz.

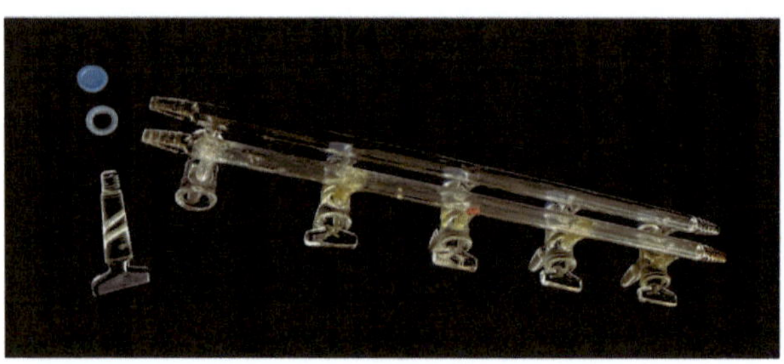

Figure 8.10 Close-up view of a Schlenk double manifold, with its double-oblique stopcocks.

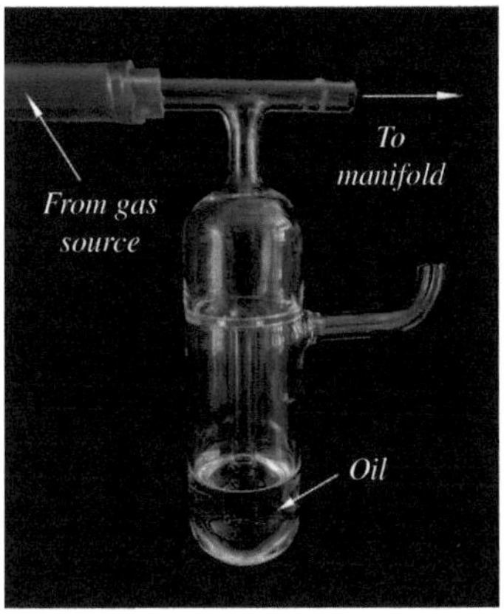

Figure 8.11 Bubbler close-up view.

a bubbler with an "anti-suck-back" valve (Figure 8.11), and a pressure-regulated gas source (from a cylinder or an outlet). The double manifold has "double oblique" stopcocks (see Figure 8.10) that can deliver slightly overpressurized inert gas in one position and vacuum in the 180° rotated position.

With a Schlenk line up and running, you can connect a flask to one of the manifold outlets and evacuate by setting the stopcock in the "vacuum" position. The evacuated flask can then quickly be refilled with inert gas by turning the stopcock 180° to the "inert gas" position. Very practical!

There are many uses of a Schlenk line, see, for example, Section 2.6 of Chapter 2 and/or the excellent review by Borys [2].

8.3 Heat Sources

8.3.1 General

Heat sources are indispensable in organic chemistry labs. The iconic Bunsen burner (Figure 8.12) is still used in some situations. Historically, municipal gas was used for this burner, but nowadays propane is the most common. A very practical solution, often seen in smaller labs, is to use a simple propane-powered camping burner (Figure 8.13).

Heat guns (Figure 8.13) are used for purposes that do not require the heat of an open flame (e.g., for charring TLC plates).

For heating baths, a "magnetic stirrer hotplate" (Figure 8.13) is the standard, see also Section 4.5.2 of Chapter 4 and Section 8.4 in this chapter.

An electric heating mantle (see Figure 4.24 in Section 4.5.4 of Chapter 4) is an alternative to heating baths, especially for larger (< 1 L) distilling flasks.

Microwave devices are newcomers as heat sources in the lab; for more information on their uses and advances, see reviews [3, 4].

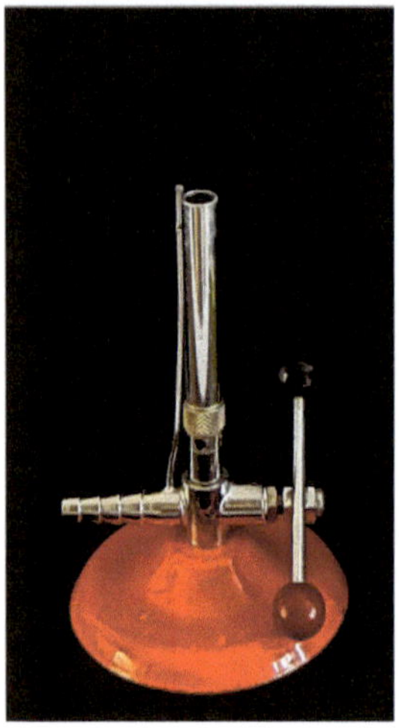

Figure 8.12 The Bunsen burner.

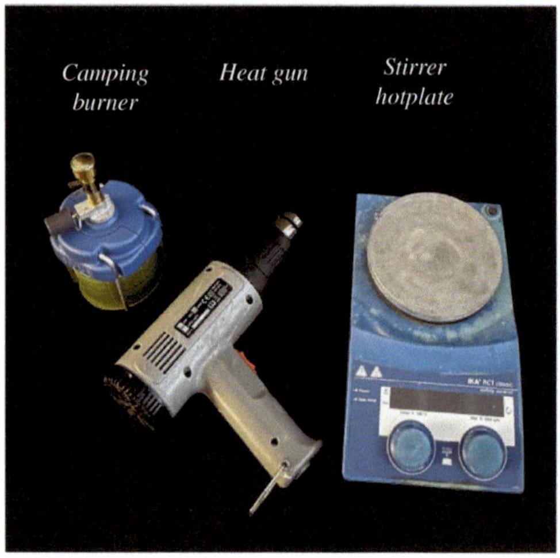

Figure 8.13 Three other common heat sources.

8.4 Stirring Equipment

8.4.1 General

One of the most commonly used stirring devices in today's organic chemistry laboratories is the magnetic stirrer plate (Figure 8.14a). It is used for stirring liquids in any containers ranging from small sub-milliliter vials to large multi-liter flasks. The principle is simple: a flask standing on a plate holds a Teflon-coated magnet ("magnetic stirrer bar") which is moved, through magnetic contact only, by a rotating magnet underneath the plate. The rate of stirring can be adjusted. Most magnetic stirrers today can also provide heating ("magnetic stirrer hotplates").

Despite their versatility, magnetic stirrers have limitations: high-viscosity mixtures can require more force than the magnetic stirring bar can produce, with the result that the stirring bar stops even if the magnet underneath still rotates. There is also a volume issue: larger (multi-liter) solutions will require large, heavy stirring bars that can actually cause breakup accidents if they start wobbling inside of a thin-walled glass vessel. Therefore, when the liquid volume exceeds a few liters, one should consider using other stirring devices, like the stirrer–rod–paddle assembly (Figure 8.14b). The difference compared to magnetic stirrers is that now the stirring "paddle" is physically connected to a rotating motor via a metal or glass rod. If the stirred solution needs to be in a closed flask, the rotating rod must pass through a sealing gasket. There are many types of such gaskets, including glass-rod-to-glass, metal-rod-to-Teflon, and glass-rod-to-Teflon, and the choice depends on how air-tight, chemically resistant, etc. the gasket needs to be.

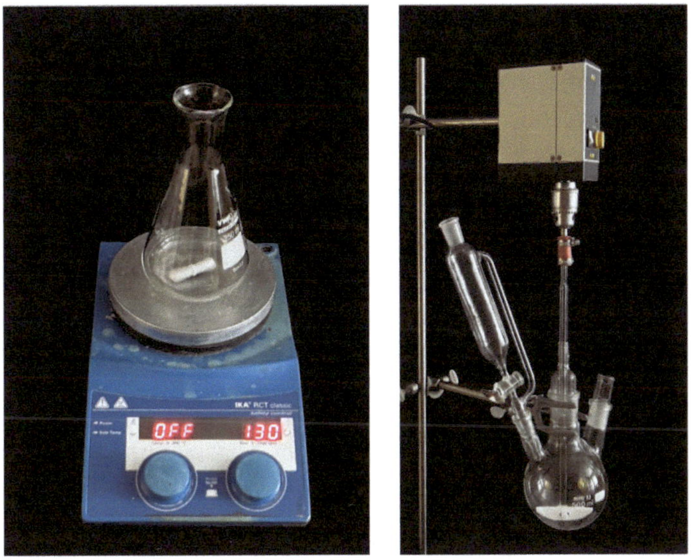

(a) (b)

Figure 8.14 (a) A magnetic stirrer hotplate. (b) A stirrer–rod–paddle assembly.

8.5 Balances

8.5.1 General

Balances (Figure 8.15a,b) are indispensable in all organic chemistry works, and a laboratory should have access to both two-decimal and three/four-decimal gram balances. With the advent of electronic balances, almost all of the maintenance and

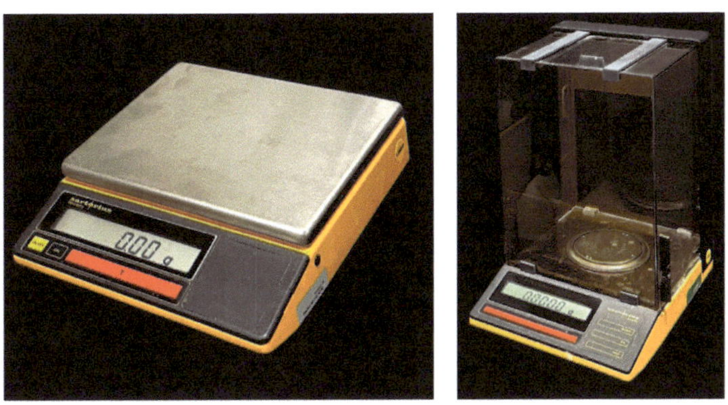

(a) (b)

Figure 8.15 (a) A two-decimal laboratory balance. (b) A four-decimal analytical balance.

calibration works have disappeared, so not much has to be said here compared to what would have been necessary a century ago when mechanical balances dominated. However, a few things should be remembered.

8.5.2 Tips and Tricks

1. *Location*: place your balances in vibration- and draft-free environments on sturdy supporting benches.
2. *Weighing*: use small beakers or small containers made of folded paper or aluminum foil for substance weighing. Try to avoid spills as much as possible. However, sooner or later spills will happen, then remember that modern electronic devices are very easy to disassemble for cleaning, so clean whenever necessary.
3. *Accuracy*: be skeptical against very accurate weight numbers, especially when weighing large glass vessels with very small amounts of material in them. The empty weight of, for example, a 250 mL round-bottomed flask can vary by several milligrams depending on the amount of moisture adsorbed on to its surface. The typical example is when you have taken a flask from its shelf, carefully weighed it with three- or four-decimal balance, and filled it with an aqueous solution with a few milligrams of your interested compound. After completing freeze-drying, the weight is slightly *less* than the empty weight! What has happened is that freeze-drying removed not only all water from your solution, but also few milligrams of moisture that was initially adsorbed on to the inner surface of your flask! In such cases, the weight should increase slightly over time when the flask readsorbs moisture from the air.

For better accuracy in cases like this, you should use glass vessels for evaporation that are not unnecessarily large in relation to the amount of expected residue. Another way, when both the solution volume and substance amounts are very tiny, is to evaporate from vessels made of plastic instead of glass, e.g., Eppendorf or Falcon tubes. Plastic is less hygroscopic than glass, and therefore adsorbs less water from the air.

8.6 Labware

8.6.1 Dishwashing, General

Your labware needs frequent care. Dishwashing is as necessary in the lab as in the kitchen. Like in the kitchen, it can be done manually or with a dishwasher. Admittedly, the procedures and equipment are different from the kitchen ones. Manual dishwashing of glassware, common in smaller labs, is done simply by rinsing with solvent (acetone, ethanol, or water) dispensed from a spray bottle (Figure 8.16), sometimes complemented with a regular aqueous brush-and-detergent wash. Final rinse is done with acetone or ethanol.

More resourceful labs use lab dishwashers. They resemble kitchen dishwashers, but use stronger detergents, higher water temperatures, and deionized water for the last rinse. Lab dishwashers are also more expensive than their kitchen counterparts.

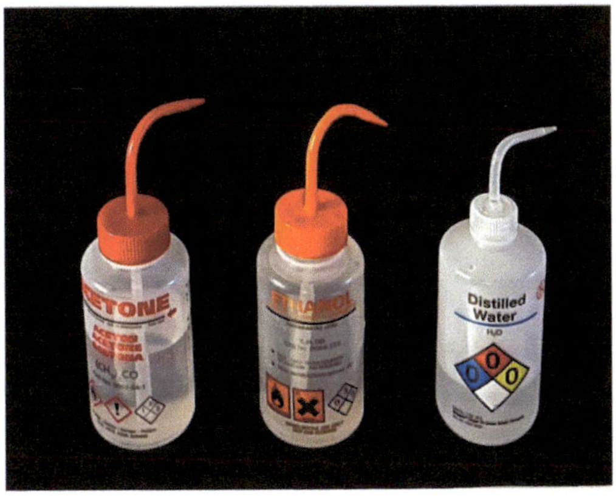

Figure 8.16 Three manual dishwashing tools: spray bottles with acetone, ethanol, and water.

A common arrangement is that all chemists in one lab room share a common dishwasher. It is also common practice to manually prerinse the glassware with organic solvents whenever it is deemed necessary. Some molecules do not dissolve even in hot water.

The most advanced dishwashing arrangement is where separate staff members do the dishwashing. They collect the dirty items in the labs, wash them in a central lab dishwasher, and redistribute the cleaned and dried items to the labs. With this arrangement, the chemists usually have restrictions as to what they can put in the dirty dishes, since poisonous and corrosive residuals are risky for the dishwashing staff (and the dishwasher). In practice, this means that the chemists *must* prerinse the glassware themselves before passing it on for dishwashing.

8.6.2 Dishwashing, Tips and Tricks

8.6.2.1 Dishwashing: Inside Glass Areas Difficult to Reach

A large vessel with a narrow opening may not allow entry of a brush or other device. If strong chemicals do not do the cleaning job alone, then some added sea sand and gentle shaking can help with the mechanical part of the cleaning. Narrow glass tubes or connections can be cleaned with a pipe cleaner.

8.6.2.2 Dishwashing: Strong Chemicals

Some glass surfaces are difficult to clean efficiently even with the washing procedures mentioned. A clogged filter glass frit is an example. If the "dirt" cannot be removed by other procedures, then strongly acidic and oxidizing mixtures can be used. One such well-known mixture is "chromic acid solution." It is a viscous, reddish-brown liquid that consists of "chromic acid" in concentrated sulfuric acid. When it is used for removing organic "dirt," the chromium is reduced, and the solution gradually turns greenish, losing efficiency.

Another, chromium-free, alternative is "piranha solution," which consists of 30% hydrogen peroxide in concentrated sulfuric acid.

Yet another alternative is "aqua regia," the classical 1:3 mixture of concentrated nitric and hydrochloric acids that can dissolve gold. Like the other two mixtures, aqua regia is both strongly acidic and strongly oxidizing.

A chemical-free (but rather expensive) alternative for removing organic deposits from glass is to perform a "bake" of dishwasher-cleaned glassware in a muffle furnace at high (4–500 °C) temperatures for several hours. This treatment is common in analytical laboratories, and ensures the absence of everything that can be burned away.

Suggestions on how to make and use chromic acid, piranha, and aqua regia solutions are found below. Note that preparation and use of these cleaning solutions is associated with *risks*. As usual, the larger the amounts, the higher the risks:

1. *How to prepare and use chromic acid cleaning solution*: caution: perform all work in a fume hood. Always use proper hand and eye protection when handling this mixture! It is *very* corrosive (that is why it is such a good cleaning agent). Chromic compounds are also carcinogenic. Check your local regulations concerning use of chromic acid and disposal procedures before you start!
2. *Preparation*: use finely powdered commercial or technical grade sodium dichromate or potassium dichromate (4 g). Add a very small amount of water, enough to make a paste. Then, slowly add 60 mL of concentrated sulfuric acid while stirring well with a glass rod (caution: heat is released!). Keep the mixture in a glass-stoppered bottle. Larger volumes can be prepared but be restrictive because of the toxic and corrosive properties.
3. *Use*: the glassware should be precleaned and dry with no solvent residues! Fill with the cleaning solution and let it work as long as necessary. Empty slowly (possible heat evolution!) into a dedicated waste container prefilled with excess water, and rinse several times with distilled water. Alternatively, make a *small* bath with the cleaning solution and bathe the glassware. Rinse as above. *Disposal of waste container contents*: follow local regulations!
4. *Disposal of mixture not used*: slowly add the mixture to excess ice-slush, *not* the other way! Neutralize, and dispose of in compliance with local regulations.
5. *Note*: glassware cleaned with chromic acid solution retains trace amounts of surface-bound Cr^{3+} ions, even after repeated water washings. This makes chromic acid *unsuitable* for cleaning of NMR tubes (Cr^{3+} ions are paramagnetic).
6. *How to prepare and use piranha mixture*: before even considering the piranha mixture, read the review about it [5]. Of the three mixtures described here, piranha is the one with the highest risk, it can explode when handled incorrectly. It should therefore not be stored but always used directly after preparation and then be properly disposed of immediately afterward. In the lab, piranha is mainly used for final cleaning of cuvettes, NMR tubes, or small glass frits. Volumes larger than 20 mL are seldom needed and should be avoided. *Curiosity*: other than in the lab, piranha mixtures are used for etching during semiconductor fabrication.
7. *Preparation (20 mL example)*: always use proper hand and eye protection and work in a hood with the window between you and the mixture. Add 17 mL of pure concentrated sulfuric acid to a beaker cooled in an ice bath, then stir while slowly adding 3 mL of 30% hydrogen peroxide. Remove the ice bath and let the mixture stabilize for 30 minutes at room temperature.

8. *Use*: the glassware should be *precleaned and dry* (no solvent residues!). Still working in a hood, transfer the piranha solution to the glassware to be cleaned with a glass pipette or similar, and let it work for 5–20 minutes. Then empty slowly (possible heat evolution!) into a dedicated waste container prefilled with excess water, and rinse several times with distilled water. Rinse the glassware with plenty of distilled water.
9. *Disposal (as soon as possible after use)*: slowly pour the entire waste container contents into a stirred beaker filled with excess ice-slush. If required, neutralize with 1 M aqueous sodium hydroxide. Dispose in compliance with local regulations.
10. *Final note*: stabilized mixtures related to piranha are commercially available in some countries. Consult your local vendor.
11. *How to prepare and use aqua regia*: as with piranha, this corrosive mixture should not be stored but always be used directly after preparation and then be disposed of afterward. Since toxic fumes can be released during preparation and use, always work in a well-ventilated hood and use proper eye and hand protection. In the lab, aqua regia is mainly used for final cleaning of cuvettes, NMR tubes, or small glass frits. Volumes larger than 50 mL are seldom needed and should be avoided.
12. *Preparation*: slowly add one part of 68–70% HNO_3 to three parts of 30–35% HCl while stirring continuously. Be cautious, as the reaction can generate heat and toxic fumes. Let the mixture stabilize a few minutes before use.
13. *Use*: the aqua regia should not be "older" than 24 hours. The glassware should be precleaned and dry (no solvent residues!). Still working in a hood, transfer the aqua regia to the glassware to be cleaned with a glass pipette or similar, and let it work for 5–20 minutes. Then empty slowly (possible heat evolution!) into a dedicated waste container prefilled with excess water, and rinse several times with distilled water. Rinse the glassware with plenty of distilled water.
14. *Disposal (as soon as possible after use)*: slowly pour the entire waste container contents into a stirred beaker filled with excess ice-slush. If required, neutralize with 1 M aqueous sodium hydroxide. Dispose in compliance with local regulations.

8.6.2.3 Dryness

Glass surfaces can hold small amounts of moisture from the air even if they look perfectly dry. This causes well-known problems with, for example, organomagnesium or organolithium reagents. Read more about moisture in Section 2.5 of Chapter 2.

Apart from its influence on certain reactions, adsorbed moisture can cause problems when determining the weight of the residual substance in glass flasks after solvent removal. Subtraction of the flask's tare weight from the flask's filled weight gives an underestimated value, especially after a freeze-drying operation. The reason is that more adsorbed moisture is present on the inside glass surface during tare weighing than directly after freeze-drying. The difference in weight can be several milligrams for a typical 250 mL round-bottomed flask.

The lesson learned from this is to avoid oversized glass flasks if you want to weigh freeze-dried material with any accuracy. The error from adsorbed water becomes less significant when the ratio of solvent amount (which determines the flask size) to the residual substance amount is smaller. Another solution is to freeze-dry from a flask or vial made of plastic, which adsorbs less moisture than glass.

8.6.3 Glass Joints, Sealing, Stoppers, and Septa

8.6.3.1 Glass Joints and Sealings

Most glassware today have ground glass joints with standard sizes, like NS 14/23, NS 19/26, NS 24/29, and 29/32. The two "NS" ("Normal Schliff") numbers refer to the widest outer diameter and the length of the ground glass section, respectively. To be "leakproof," ground glass joints need some type of sealant. The sealant also prevents "joint freezing," a common problem (see Section 8.6.4). Stopcock grease has long been the standard sealant, and several variants are available. The Teflon sleeve is an alternative to grease that can be used when grease contamination must be avoided.

8.6.3.2 Stoppers and Septa

A cork or rubber stopper was used to seal bottles and flasks before the ground glass stopper became standard. For small-scale work under inert gas, a rubber septum has nowadays become a popular alternative (see Figure 8.17). When used correctly, a rubber septum on a reaction flask allows for easy and leak-proof addition or withdrawal of liquid with a syringe/needle. However, there are some risks associated with

Figure 8.17 Glass stoppers and septum.

the use of septa: (i) rubber is not fully compatible with some solvents (e.g., chloroform), so care must be taken to avoid repeated direct contact. (ii) Repeated needle penetration in the same location on the septum increases the risk of leakage. Always shift the needle location a little for every new penetration. (iii) Never reuse septa, always use fresh ones for a setup! Rubber ages and becomes brittle rather quickly in air.

8.6.4 Glass Joints, Tips and Tricks

Frozen glass joints: almost every chemist has encountered this problem, often during glassware dishwashing. No matter how you try, two joined items (often a glass stopper in a flask) refuse to separate, and the situation can end with broken glass or even bloodshed.

1. when applying force with your hands to any type of glassware, use a towel or similar between your hands and the glass. This reduces the risk of getting cut if the glass breaks.
2. warm the outside (all around and quickly!) of a resistant glass joint with a flame or heat gun. This expands the outside and can lead to a release if the joint is twisted immediately after warming. Be sure to use a towel here too when you twist!
3. gentle knocking with a tree club or similar in the right direction can sometimes do the trick.
4. glass joints often get frozen because aqueous alkali has been used and remains as a "glue" in the joint. It could then help to pipette some drops of concentrated hydrochloric acid where the joint ends. The acid will seep in between the stuck glass surfaces, and if it is left to do its work overnight, it might have dissolved enough alkali to make the joint separable again. If the problem is instead a remaining organic solid, it might help to use a few drops of chloroform or acetone in the same way.
5. a few minutes in an ultrasound bath can solve the frozen joint problem in some cases.

8.7 Storage of Chemicals

8.7.1 General

Every chemistry lab needs to store chemicals, temporarily and/or more permanently. The storage requirements differ depending on the properties of the particular chemical. Ordinary wall or cupboard shelves are suitable for nonvolatile, nontoxic, nonhygroscopic, and inflammable compounds, whereas most other categories need special arrangements. Below are some recommendations.

8.7.2 Solvents, Both Flammable and Inflammable

Bulk solvent containers (1–2.5 L) are best stored in fire-withstanding metal cabinets (Figure 8.18), even if they are not flammable. A separate room is preferable.

Figure 8.18 Metal cabinet for solvents.

Smaller "every-day-use" bottles of solvents can be stored in a ventilated space near each workplace (local regulations often specify the maximum allowable amounts of flammable solvents in a room). For examples, see Chapter 7.

8.7.3 Toxic, Corrosive, and/or Volatile Compounds

These are best stored permanently in separate rooms. Most labs have such rooms where they store hundreds of small (1–500 g) bottles and cans in ventilated cabinets

(a) (b)

Figure 8.19 (a) A ventilated cabinet. (b) Removable plastic boxes that subdivide and protect shelve content.

(Figure 8.19a). A very practical solution is if the cabinet shelves are further divided into smaller, removable plastic boxes that contain 10–20 bottles (Figure 8.19b). This simplifies searching and also reduces the damage if a bottle should accidentally start leaking.

Smaller "every-day-use" bottles can be stored near each workplace in a ventilated space. For examples, see Figure 7.4 in Section 7.3 of Chapter 7.

8.7.4 Severely Toxic Compounds and Controlled Substances

Severely toxic compounds (labeled with the skull pictogram, Figure 8.20a) or controlled narcotic substances should be stored separately in a locked metal cabinet (Figure 8.20b). The rules differ between countries, so check your local regulations.

(a) (b)

Figure 8.20 (a) The skull pictogram. (b) A locked cabinet for severely toxic compounds.

(a) (b)

Figure 8.21 (a) Desiccator. (b) Dry cabinet.

8.7.5 Hygroscopic Compounds

These should be stored in a dry space, such as a desiccator or a dry plexiglass cabinet (see Figure 8.21a,b). The desiccant inside (blue) could be silica gel, calcium sulphate (Drierite), calcium chloride, or phosphorous pentoxide (a very strong desiccant!). If the compounds are not volatile, a desiccator could be evacuated *if* (and only if!) the compound container lids are loosened so that gas can pass in and out.

If hygroscopic compounds are in the form of liquids under dry inert gas in a septum-sealed bottle, they can be stored in an ordinary ventilated cabinet or in a fridge/freezer.

8.7.6 Air-sensitive Compounds

These could be stored as hygroscopic compounds, but with the added requirement that they need constant inert gas blanketing.

8.7.7 Temperature-sensitive Compounds

These should be stored according to the label recommendations in either a fridge (0–4 °C) or a freezer (−20–30 °C). To minimize the risk of contamination of these closed storage spaces, every flask and bottle should be sealed with a reliable cap. Leakage of vapors in fridges or freezers can lead to toxic contaminations or, if the vapors are flammable, explosions. To further reduce the risk of explosions, the fridge/freezer should be spark-free. Most modern devices are spark-free, but older units can have the lamp switch on the inside. This switch can generate sparks strong enough to trigger an explosion if flammable vapors are present inside.

8.7.8 Gas Tubes

Local regulations for storing gas tubes should be consulted, they may differ depending on both size and type of gas. As a minimum requirement in the lab, gas tubes should be secured against tipping with chains or similar. You would not want a heavy gas tube tipping onto your toes! Damage to the outlet valve may also cause a violent release of the highly pressurized gas content.

8.7.9 Store-keeping

Finally, something should be mentioned about "store-keeping" of chemicals. The need for this is great in a lab with perhaps several hundreds of bottles, and that is for many reasons. One reason is that local laws and regulations may require a full digital registration of all chemicals present in a lab; another is that the working chemists need a fast way of finding where and whether a certain chemical can be found in the lab.

Large organizations often have their own specialized software and database to keep track of all their chemicals. Smaller labs can either use commercial software specialized for creating and managing "chemical" databases, or simply create their own database with a common spreadsheet program like Excel.

Regardless of in which database you have registered your chemicals, it must also be easy to locate them, physically (in which room and in which cabinet?). The level of detail provided in a database is usually not too high (room and cabinet, no more). The way that the chemicals in one cabinet are ordered is therefore important for quick localization of a specific bottle. Some labs use an alphabetical shelf order, based on the chemical's name in a certain language (usually English). Other labs use a language-independent shelf order based on the sum formula (best for organic compounds). The chemicals are first grouped according to their amount of carbons, and then subgrouped according to their amount of hydrogens (see Figure 8.19b). For inorganic compounds, an element-based order can be used in a similar way.

Bottles and cans stored in a well-defined order, an associated database with properly searchable fields, and a well-established routine for database management (periodical entry and removal of items) are invaluable assets for a lab. They minimize the time for running around searching for chemicals and also eliminate unnecessary purchases. Finally, they minimize the risk of getting into trouble with authorities for lack of adherence to local regulations.

8.8 Setting Up an Organic Chemistry Lab: Equipment and Analytical Instruments

8.8.1 Equipment

If you are a newly employed PhD student or postdoc, you are usually presented with a more or less complete organic chemistry lab to begin work in. However, over time, you will probably develop your own ideas on how to improve things in the lab by acquiring extra equipment or rearrange lab functions in a better way. If so, this book or its references may be of some use to you.

In the rarer situation that you will have to build up a lab from scratch, you *must* rely on your own ideas on how a lab should be organized. Perhaps you can draw from earlier workplace experience, but you will quickly realize that memory is short and you may need inspiration from existing labs, or from checking what the literature can offer. It turns out that you can find rather detailed literature descriptions of what is needed in an organic chemistry lab, even lists of the most necessary items down to the flask level. Therefore, we will here mostly refer to existing information [6–8] and not go into details with every piece of "small" lab equipment. Rotavapors and freeze-driers, however, are given special attention in Sections 4.4 and 4.6 of Chapter 4. Some information about vacuum pumps, stirrers, and balances can also be found in Sections 8.2, 8.4, and 8.5 of this chapter. In the rest of the chapter, we will discuss the types of *analytical instruments* that the organic chemist may need (primarily NMR and MS).

8.8.2 Analytical Instruments

1. *NMR and LC-MS (Figure 8.22a,b, respectively)*: For the average organic chemist in the lab, the most important analytical instruments are the NMR and LC–MS instruments. Easy access is essential and most labs, big or small, try their best to provide it. However, most chemists have only a vague understanding of the financial and maintenance resources that these instruments require, which far exceed what are required for any other piece of lab equipment. Especially, NMR spectrometers are outstanding. The price for a low-end, "budget" permanent magnet instrument starts at $50,000, and a 300 MHz superconducting instrument starts at $150,000. In addition to the purchase price, the lab also has to cover yearly costs for service, maintenance, and, in the case of a superconducting NMR instrument, the cost of liquid helium and nitrogen ($5–10,000/year).

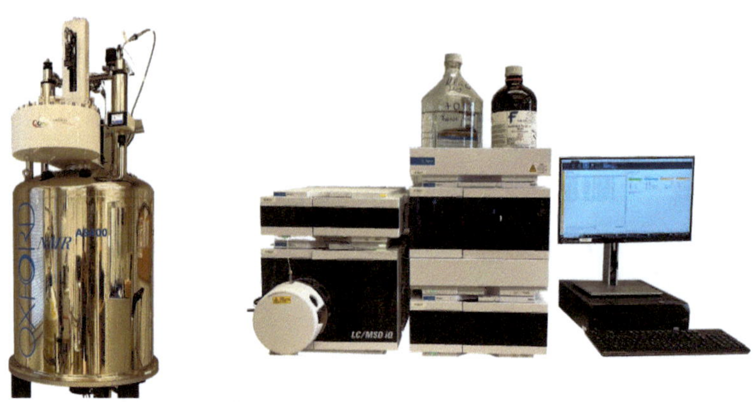

(a) (b)

Figure 8.22 (a) NMR instrument (superconducting magnet). (b) LC–MS instrument.

Larger academic departments and large company labs do have the financial resources needed for in-house NMR instruments and usually possess one or several superconducting instruments. However, smaller educational institutions and small company labs normally cannot afford an NMR instrument. Instead, they rely on contract providers to run their NMR spectra. The latter solution means less frequent access to an instrument, especially in the case of a pay-per-spectrum arrangement. This makes the (otherwise common) use of NMR for identifying synthetic intermediates less practical. The contract NMR services are used only when they are absolutely necessary, e.g., when isomers have to be distinguished or final products characterized.

The identifications of reaction products and synthetic intermediates in labs without an in-house NMR instrument are instead often performed with an LC–MS instrument. In a multistep organic synthesis, a quick verification of a reaction product's molecular weight is often all that is needed to proceed to the next step. LC–MS instruments are less expensive to purchase and service (costs are not anywhere near to those of NMR instruments) and there is a large second-hand market, which can further reduce the purchase price. You can often find a fully functional LC–MS instrument for less than $20,000.

2. *High-resolution mass spectrometer*: literature reports of new synthetic compounds should contain a high-resolution mass spectrum (HRMS) showing the mass of a mononuclidic molecular ion adduct with four decimal places. From this mass, it is often possible to extract a sum formula for the new compound. The information is not only needed for publication purposes, but also useful for rapid identification of isolated natural products.

   An HRMS mass spectrometer is a major investment, and since preparations/discoveries of new compounds are not everyday events, most organic chemistry labs outsource their acquiring of HRMSs.
3. *Other analytical instruments*: organic chemists need several other instruments than the ones mentioned above but much less frequently, e.g., an IR spectrometer, an ordinary analytical HPLC, a polarimeter (Section 5.4 of Chapter 5), a melting point apparatus (Section 5.5 of Chapter 5), and perhaps also a refractometer. Depending on the focus of the lab, these instruments may or may not be needed in-house. None of them are nearly as expensive as NMR, LC–MS, or HRMS instruments.

References

1 Cubberley, M.S., W.A. Hess, and M.B. Johnson, An inexpensive recirculating water vacuum pump for the chemistry laboratory. J Chem Educ, 2020. 97: pp. 1495–1499.

2 Borys, A.M., An illustrated guide to Schlenk line techniques. Organometallics, 2023. 42: pp. 182–196.

3 de la Hoz, A., Á. Díaz-Ortiz, and A. Moreno, Microwaves in organic synthesis. Thermal and non-thermal microwave effects. Chem Soc Rev, 2005. 34(2): pp. 164–178.

4 Abramovitch, R.A., Applications of microwave energy in organic chemistry. A review. Org Prep Proced Int, 1991. 23(6): pp. 683–711.
5 Schmidt, H.G., Safe piranhas: a review of methods and protocols. ACS Chem Health Saf, 2022. 29: pp. 54–61.
6 Leonard, J., B. Lygo, and G. Procter, *Advanced Practical Organic Chemistry*, 3rd Ed. 2013, Boca Raton, FL: CRC Press. 356 pages.
7 Vogel, A.I., et al., *Vogel's Textbook of Practical Organic Chemistry*, 5th Ed. 1989, Essex, England: Longman Scientific & Technical and John Wiley & Sons. 1514 pages.
8 Keese, R., M.P. Brändle, and T.P. Toube, *Practical Organic Synthesis: A Student's Guide*. 2006, Chichester, UK: John Wiley & Sons. 208 pages.

9

Multistep Organic Synthesis

Very complex molecules can be synthesized from simple starting materials by *multistep organic synthesis*. Each step is one chemical reaction, often complete with setup/running, monitoring, workup, and characterization. The scheme in Figure 9.1 shows a synthesis of the antiviral drug oseltamivir, published in 2006 by Corey and colleagues [1]. We have already seen one step (encircled, 5 → 6) in this synthesis as an example in Chapter 1. We will now take a closer look at the entire 12-step reaction sequence.

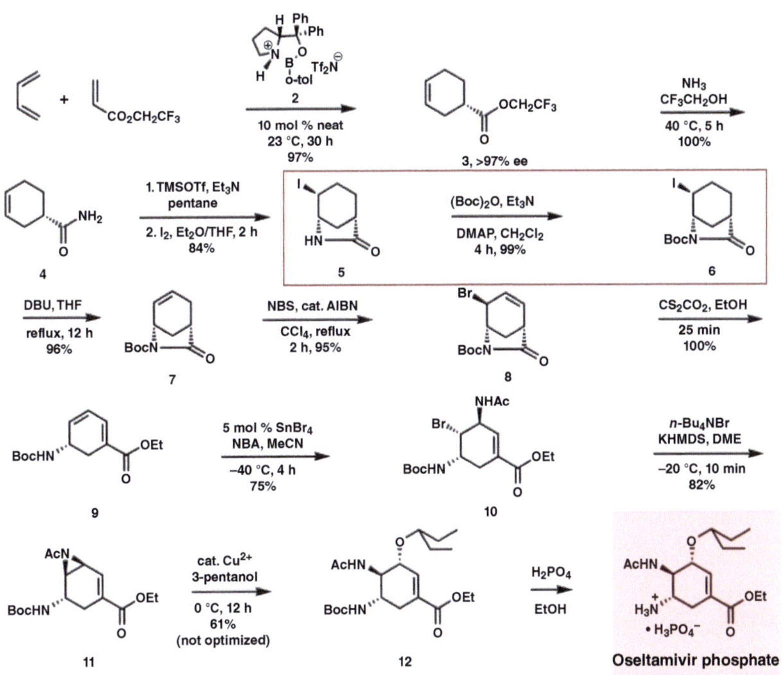

Figure 9.1 A synthetic route to oseltamivir phosphate. The encircled reaction is the acylation used in Chapter 1 to exemplify one synthetic step. *Source*: [1] / with permission of American Chemical Society.

As can be seen in the scheme (Figure 9.1), the core carbon skeleton of oseltamivir was formed already in the first step. The remaining steps placed the functional groups in their correct positions and with the correct stereochemistry.

The experimental details of the entire synthesis include eight "cold" reactions, two "room-temperature" reactions, and two "hot" reactions. As a general precaution, all reactions were conducted in flame-dried glassware and under inert gas. All reactions were monitored by TLC.

The workup procedures included five extractions, twelve evaporations, seven chromatographic purifications (silica gel columns), and four filtrations.

Characterizations were by a combination of H- and C-NMR (ten compounds), IR (ten compounds), MS or HRMS (seven compounds), optical rotation (nine compounds), and melting point (six compounds).

Obviously, a multistep synthesis like this one involves *a lot* of work, both theoretical and practical. First, a suitable synthetic route has to be designed "on paper," using a suitable sequence of known reactions. The number of known reactions is immense and the combination possibilities are almost endless, so this is a task that requires *extensive* experience.

Then the synthetic route must be tested in practice. In the majority of cases, the original route has to be modified (or even abandoned) because one or more steps does not work as expected. However, after much (it can take years!) experimentation, a successful synthetic route, leading to the desired molecule, can eventually be found. Overall, synthesis planning/design and testing is a high-level intellectual and practical exercise that requires experience, creativity, and patience. The "star" scientists of the field are well-known celebrities (some are Nobel laureates), and many review articles and books have been written about various aspects of multistep organic synthesis [2–7].

In light of the above, it is not surprising that many labs have tried to use modern computer technology for assistance in the first part of the process, synthesis planning/design [8]. Several software packages that use large reaction databases are available, and we can expect the rapid development of artificial intelligence to further influence the planning/design of multistep organic synthesis [8, 9]. However, and this is my personal opinion, the assessment of the practical feasibility of a certain organic reaction in a synthetic sequence cannot be entirely left to computers, and surely not the experimental testing of the entire synthetic sequence. These activities *will* require organic chemists with extensive laboratory experience. It is to aid in accumulation of such experience that this book was put together.

References

1 Yeung, Y.-Y., S. Hong, and E.J. Corey, A short enantioselective pathway for the synthesis of the anti-influenza neuramidase inhibitor oseltamivir from 1,3-butadiene and acrylic acid. J Am Chem Soc, 2006. 128: pp. 6310–6311.

2 Bindra, J.S., S. Ranganathan, and N. Anand, *Art in Organic Synthesis*, 2nd Ed. 1988, New York, NY: Wiley-Interscience. 448 pages.

3 Nicolaou, K.C. and E.J. Sorensen, *Classics in Total Synthesis: Targets, Strategies, Methods*. 1996, Weinheim: VCH Verlagsgesellschaft mbH; New York: VCH Publishers, Inc. 821 pages.
4 Nicolaou, K.C. and S.A. Snyder, *Classics in Total Synthesis II: More Targets, Strategies, Methods*. 2003, Weinheim: Wiley-VCH Verlag GmbH & Co. KGaA. 658 pages.
5 Nicolaou, K.C. and J.S. Chen, *Classics in Total Synthesis III: Further Targets, Strategies, Methods*. 2011, Weinheim: Wiley-VCH Verlag GmbH & Co. KGaA. 770 pages.
6 Hoffmann, R.W., *Elements of Synthesis Planning*. 2009, Heidelberg: Springer-Verlag. 227 pages.
7 Warren, S. and P. Wyatt, *Organic Synthesis: The Disconnection Approach*, 2nd Ed. 2008, Chichester, UK: John Wiley & Sons. 352 pages.
8 Feng, F., L. Lai, and J. Pei, Computational chemical synthesis analysis and pathway design. Front Chem, 2018. 6: pp. 199: 1–199:10.
9 Ali, R.S.A.E., et al., Machine learning advancements in organic synthesis: A focused exploration of artificial intelligence applications in chemistry. Artif Intell Chem, 2024. 2(1): p. 100049.

Index

Note: Page numbers in *italics* indicate figures.